# THE FRONTIERS COLLECTION

*Series Editors*

Avshalom C. Elitzur
Unit of Interdisciplinary Studies, Bar-Ilan University, 52900, Ramat-Gan, Israel
e-mail: avshalom.elitzur@weizmann.ac.il

Laura Mersini-Houghton
Department of Physics, University of North Carolina, Chapel Hill, NC 27599-3255
USA
e-mail: mersini@physics.unc.edu

Maximilian Schlosshauer
Department of Physics, University of Portland,
5000 North Willamette Boulevard Portland, OR 97203, USA
e-mail: schlossh@up.edu

Mark P. Silverman
Department of Physics, Trinity College, Hartford, CT 06106, USA
e-mail: mark.silverman@trincoll.edu

Jack A. Tuszynski
Department of Physics, University of Alberta, Edmonton, AB T6G 1Z2, Canada
e-mail: jtus@phys.ualberta.ca

Rudy Vaas
Center for Philosophy and Foundations of Science, University of Giessen, 35394,
Giessen, Germany
e-mail: ruediger.vaas@t-online.de

H. Dieter Zeh
Gaiberger Straße 38, 69151, Waldhilsbach, Germany
e-mail: zeh@uni-heidelberg.de

For further volumes:
http://www.springer.com/series/5342

# THE FRONTIERS COLLECTION

The books in this collection are devoted to challenging and open problems at the forefront of modern science, including related philosophical debates. In contrast to typical research monographs, however, they strive to present their topics in a manner accessible also to scientifically literate non-specialists wishing to gain insight into the deeper implications and fascinating questions involved. Taken as a whole, the series reflects the need for a fundamental and interdisciplinary approach to modern science. Furthermore, it is intended to encourage active scientists in all areas to ponder over important and perhaps controversial issues beyond their own speciality. Extending from quantum physics and relativity to entropy, consciousness and complex systems—the Frontiers Collection will inspire readers to push back the frontiers of their own knowledge.

For a full list of published titles, please see back of book or springer.com/series/5342

Gennadiy Zhegunov

# THE DUAL NATURE OF LIFE

## Interplay of the Individual and the Genome

Translated from Russian by
Denys Pogozhykh and Iryna Ashby

Edited by Eddie Kalmykov and Denys Pogozhykh

 Springer

Prof. Dr. Gennadiy Zhegunov
Kharkov State Veterinary Academy
Ukraine

ISSN 1612-3018
ISBN 978-3-642-43133-3        ISBN 978-3-642-30394-4 (eBook)
DOI 10.1007/978-3-642-30394-4
Springer Heidelberg New York Dordrecht London

# Preface

What is life? Although the origins and nature of life are not yet fully understood, it is a widespread phenomenon on our planet which can be characterized by a number of fundamental attributes that differentiate living bodies from non-living ones in significant ways. It is clear that life is a qualitatively specific manifestation of organized matter. However, due to the fact that all life forms are complex and diverse, there is no complete or clear understanding of this phenomenon or its prevalence.

We will attempt to analyse the mysteries of life from a new perspective. Though the material is based on the classical scientific conception of life, we will consider biological phenomena from a non-standard point of view, developing, among others, the notion of the many *dualities of life*.

In particular, when analysing the nature of life, we proceed from the fact that there are individual and specific physical beings (bodies) which, with all of their characteristics, properties, and functions, represent one manifestation of life in the natural world. On the other hand, life is also a systematic determinant of the iterative interactions of individual organisms leading to the establishment of the living world as an entity (as a global phenomenon of life), and therefore also requires scientific and philosophical comprehension on a global scale.

This duality is also tied to the fact that organisms are physical and visual (phenomic) derivatives of the expression of their genomes. This book suggests and develops a hypothesis about the primary nature of genomes in relation to cells and the living bodies they comprise. Life, in other words, includes complex visible phenotypical components and even more complex invisible genotypical ones. Because this duality is present not only at the level of the individual organism but also in the combined existence of global life, we will attempt to understand the concept of the global phenome, which is a result of the expression of the global genome. A final duality is also seen in the immortality of the global phenomenon of life that is maintained by the mortality and replacement of its discrete components, on both an individual and a global level. It may therefore be possible to surmise that this duality is the essence of evolution, by asking whether it is an effect of the survival of organisms and species or the survival of information, which is transferred and expressed by them.

Because life is a highly complex phenomenon, it is analysed chapter-wise from several points of view. The first part deals with life as a phenomenon of the material world, while the second explores life as the existence of *living bodies* with their inherent properties. Part 3 describes various *processes and mechanisms* that are usually ignored or mixed in with principally different categories. The fourth part delves into the role of *information*, which is a *directive and creative force* of all life processes and mechanisms. The fifth part gives a detailed analysis of the *dualism* of life, followed by a short conclusion that provides a review of the material and gives generalizing summaries.

As a whole, the book provides several key concepts regarding the nature of life. The nature or individual characteristics of humans are not greatly emphasized in the present work, because humans are considered here to be a representative of mammals without any distinct peculiarities which would require a separate and detailed analysis.

# Key Concepts

In order to follow and understand the ideas presented in the book, it is necessary to introduce several key concepts that will be used throughout:

1. *Life*. An abstract, collective term that segregates a part of Nature with specific biological properties.

2. *Phenomenon of Life*. One of the forms of organized matter. The process of continuous coexistence of all living bodies which are complementary to the diverse nature of the Earth.

3. *Living Bodies*. A form of manifestation of life. The physical bodies that possess biological properties and peculiarities. These are discrete units of life, which are temporary dissipative self-replicating biological systems. The basis of their rebirth and existence is the relatively permanent genome.

4. *Life of a body* is the dynamic process of the limited existence of discrete units of life from appearance until end.

5. *Gene*. The unit of information that controls the genesis of a certain body characteristic. Genes are segments of NA that carry out specific functions, such as protein synthesis regulation, enhancing or suppressing the actions of other genes, and so on.

6. *Genotype*. The presence of certain genes or their totality in a given individual.

7. *Phenotype*. The presence of certain characteristics or their totality in a given individual.

8. *Genome*. The entirety of genetic information stored in the DNA (or RNA) of living beings. Also, a conceptually defined specific part of a cell that contains a select set of NA and proteins united into an integrated structural and functional

system. This system contains special genetic information as well as mechanisms and instruments of its use.

9. *Phenome*. A visual set of characteristics arising from the expression of genes and from the influence of external factors. Also, a conceptually defined part of a cell that surrounds and integrates a genome, forming a monolithic body. The phenome of a multicullular organism is represented as the combined phenomes of all of its individual cells. In other words, there are highly organized colonies of standard genomes within a combined phenotypic framework.

10. *Genotypic Life*. The totality of the operational processes of all genomic elements for existence and realization of their informational potential.

11. *Phenotypic Life*. Processes of coexistence, functioning and interaction of living bodies, including all their properties and characteristics.

12. *Global Genome System*. The totality of functioning and interacting genomes of all living bodies.

13. *Global Phenome System*. The totality of functioning and interacting phenomes of all living bodies.

14. *World System of Life or Integrated Life System*. The totality of functioning and interrelated genomes and phenomes.

# Acknowledgements

First and foremost I would like to thank Dr. Johannes M. Nitsche, who was instrumental in making the translation of this book a reality. From the first contact with Springer, he acted very ably as scientific and language consultant throughout the project, and had considerable influence on the processing and improvement of the manuscript. I wish to thank Iryna Ashby and Dr. Denys Pogozhykh, who prepared the initial translation from Russian into English. Dr. Eddie Kalmykov subsequently undertook the difficult task of meticulously editing roughly the first 60% of the manuscript. The author is deeply grateful to him for all his comments, alterations, and skill. I would like to thank Dr. Denys Pogozhykh further for helping edit the remainder of the text together with Springer's inhouse editors. I express my gratitude to Professor V.V. Egorov, Professor Z.D. Vorobets, Professor N.I. Pogozhykh, Dr. D.V. Leontyev, Dr. E.G. Zhegunova (Pogozhykh), and Dr. V.F. Kopietskiy for reading through the manuscript and making valuable remarks. Special acknowledgments to Kh. Khasbaui for drawing the figures. Finally, my sincere gratitude goes to Dr. Angela Lahee at Springer for shepherding this project from inception to fruition; the book could not have been in better hands.

# Contents

**Part I   The Phenomenon of Life. Essentials**

1   Life on Earth ........................................ 3

2   Material Basis of Life ............................... 19

3   Origin and Development of Life ....................... 37

4   Discreteness, Order, Organization, and Integrity ..... 45

5   Living Systems ....................................... 53

6   Life States .......................................... 57

**Part II   Living Bodies: Carriers of Life**

7   A Mode and a Tool of Life ............................ 65

8   Cells and Organisms .................................. 73

9   Reproduction and Individual Development ............... 81

10   Evolution ........................................... 95

11   Homeostasis and the Maintenance of Integrity ........ 113

12   Ageing and Death of Individuals ..................... 123

**Part III  How Life Works: Mechanisms and Processes**
**of Living Bodies**

**13  Bodies, Processes, Mechanisms and Interactions** . . . . . . . . . . . .  137

**14  Strategy of Biological Catalysis** . . . . . . . . . . . . . . . . . . . . . . . . .  141

**15  Strategy of Copying** . . . . . . . . . . . . . . . . . . . . . . . . . . . . . . . . .  151

**16  Strategy of Self-Organization** . . . . . . . . . . . . . . . . . . . . . . . . . .  157

**17  Strategy of Matter and Energy Transformation** . . . . . . . . . . . . .  165

**18  Cell Mechanisms** . . . . . . . . . . . . . . . . . . . . . . . . . . . . . . . . . . .  183

**19  Physiological Mechanisms** . . . . . . . . . . . . . . . . . . . . . . . . . . . . .  191

**Part IV  Mechanisms of the Invisible World of Information**

**20  Biological Information and Cybernetics** . . . . . . . . . . . . . . . . . . .  201

**21  Genetic Information** . . . . . . . . . . . . . . . . . . . . . . . . . . . . . . . . . .  219

**22  Genes and Genomes** . . . . . . . . . . . . . . . . . . . . . . . . . . . . . . . . .  229

**23  Functional Systems of Genes** . . . . . . . . . . . . . . . . . . . . . . . . . . .  241

**24  Genetic Continuity** . . . . . . . . . . . . . . . . . . . . . . . . . . . . . . . . . .  251

**Part V  Duality of Life**

**25  Body and Intelligence Duality** . . . . . . . . . . . . . . . . . . . . . . . . . .  261

**26  Genomes and Their Bodies** . . . . . . . . . . . . . . . . . . . . . . . . . . . .  263

**27  Bodies and Associated Phenomena** . . . . . . . . . . . . . . . . . . . . . .  277

**Conclusion** . . . . . . . . . . . . . . . . . . . . . . . . . . . . . . . . . . . . . . . . . . . .  287

**Index** . . . . . . . . . . . . . . . . . . . . . . . . . . . . . . . . . . . . . . . . . . . . . . . .  291

**Titles in this Series** . . . . . . . . . . . . . . . . . . . . . . . . . . . . . . . . . . . . .  295

# Part I
# The Phenomenon of Life. Essentials

# Chapter 1
# Life on Earth

## 1.1 Key Features of Life

Currently, the Earth is the only stellar body known to contain life as we have defined it (see Fig. 1.1). Despite the fact that life on Earth is a unique phenomenon in our Solar System, it is very widespread and complexly diverse on this planet. It appeared at a definite stage during the Earth's development, approximately 3.5 billion years ago, as a result of spontaneous chemical interactions that led to unique organizations of matter.

Life has a cellular basis and is extremely widely distributed over the entire planet, despite dramatically varied physical and chemical conditions in the external environment. As a result of adaptation and evolution, living beings are very diverse in terms of their size, structural complexity, multicellularity, level of organization, features of metabolism, and vital functions. Such variability allows them to occupy practically any ecological niche on Earth. Living organisms can live above, in, and under the ground, or in water, air, rocks, other organisms, and in extreme conditions such as in ice or hot geysers. They can be found under enormous pressures many kilometres beneath the oceans and at very high altitudes in the anoxic atmosphere at the edge of space. Manifestations of life can also be observed at extremely low temperatures ($-50$ °C), and at very high temperatures (up to $+100$ °C). For example, some molds and fungi (*Aspergillus*, *Cladosporium*, *Helmintosporium*) are known to live on the cooling covers of nuclear reactors, surviving colossal doses of radiation.

One can say that the Earth is simply 'contaminated' with so much life that practically nothing can destroy it on our planet. The building blocks of living organisms are nucleic acid molecules (NA) and proteins, the properties and functions of which (in an aqueous environment) account for life's immense diversity. Only a catastrophe of cosmic proportions could annihilate life, such as a cataclysmic event that causes the temperature on the Earth's surface to rise above $+100$ °C leading to the disappearance of water. Man, despite his global impact on animate and inanimate Nature, cannot destroy the entirety of life. Even nuclear

G. Zhegunov, *The Dual Nature of Life*, The Frontiers Collection,
DOI: 10.1007/978-3-642-30394-4_1, © Springer-Verlag Berlin Heidelberg 2012

**Fig. 1.1** Is there life on other planets? in our solar system, life prospers only on planet Earth which is represented as the component that serves as life's primary constituent: a single cell

war and the consequent nuclear winter could destroy only 'intelligent' life and an indeterminate number of other different types of organisms. Nevertheless, many viruses, microorganisms, and more complex life forms inhabiting the depths of the Earth or the oceans would survive this 'local catastrophe' almost without any problems. Thus, life as a qualitatively special form of matter will exist as long as the conditions required for the formation of nucleic acids and proteins in aqueous solutions remain on Earth.

The fact that life exists only on Earth is due to the Earth's specific location in the Solar System and its size. Only at its current position and distance from the Sun are the temperature and other physical conditions needed for the existence of organic substances, as well as liquid water, optimal. These factors in turn make it possible for 'nucleic-protein bodies' to exist. The size of the Earth is also ideal for gravitational retention of the atmosphere, which ensures the relative stability of the physicochemical surroundings of living organisms. The atmosphere is not only an 'umbrella' against damaging ultraviolet radiation and meteorite bombardment, but it also helps to maintain the Earth's surface temperature between 0 and 100 °C. Water remains liquid (but not solid or gaseous) in this particular temperature range, which is a necessary condition for the manifestation of life. Although the substrates of living things—nucleic acids and proteins—can retain their vital potency at lower temperatures, even at temperatures approaching absolute zero, the commencement, manifestation, and proliferation of life nevertheless requires liquid water.

Some forms of life can be unnoticeable or even completely absent. For example, various primitive organisms can completely halt their vital processes, resulting in a state called anabiosis. Unicellular organisms, small invertebrates, spores, and seeds of plants can remain in this state for many years. Anabiosis makes possible the prolonged preservation of an organism's structure, and more importantly, the maintenance of the structural and functional state of its DNA and proteins during extreme conditions such as low temperatures and complete dehydration. This is why, upon the return of normal conditions, they can restore all the processes of their vital functions and subsequently *revive*. Viruses are also capable of staying in anabiosis for a long time while maintaining the integrity of their DNA, and only manifest their viral properties after entering a host cell. These facts do not eliminate, in principle, the possibility of the existence of 'latent life' on other cosmic bodies (in the form of stable nucleic acids), nor the possibility of Earth's colonization by such beings billions of years ago.

The life of any given organism is a finite, unidirectional process. A process is a course for the development of some phenomenon, a successive change of developmental states and stages. In this case, it is the process of the appearance, development, and extinction of previous generations of individuals and the formation of new ones. The process of life has one-way directionality—from the past, through the present, to the future. This is implemented in the irreversible phase changes of ontogenesis intrinsic to every living being. Generations of organisms are continuously replaced by others. The alternation of generations is a striking phenomenon that is characteristic only of communities of living organisms, and it is an amazing property peculiar to life. It started from the moment of the origin of living bodies and it continues to this day with no tangible end in sight. Life, therefore, is a continuous process, because in the course of reproduction, eternal genomes are transmitted—with slight changes—from one mortal body to the next, and from one generation to another, over millions of years and billions of generations.

Every organism as a carrier of life exists only as a constituent of an ecosystem and its environment. Only in the organism–environment system does the redistribution of matter and energy take place. That is why it is necessary to consider living matter and the sphere of its existence as a large, integrated system. Based on the ongoing processes of constant redistribution of energy and matter within such a system, life can be considered not so much as the existence of autonomous organisms, but rather as a planetary system in which these organisms are just constituents.

Life also has an informational basis, since reproduction, development, function, and evolution are based on info-genetic processes. In particular, throughout their transient existence, individuals transmit a genetic program of development to the next generation in the process of reproduction via the DNA of gametes (sexual propagation) or via the DNA of a body part (asexual propagation). In turn, these new individuals grow and mature on the basis of their DNA programming, and then produce their own gametes and reproduce, establishing a life cycle based on the information stored in nucleic acid sequences. This cycle continues as long as the conditions for survival and reproduction exist. The continuity of genetic material, despite the alternation of generations of individuals, is one of the most important characteristics of life. It accounts for the 'intermittent continuity' of life, despite the mortality of its individual constituents, since they have time to create a transitional form of existence embodied in the genetic material of gametes, spores, cysts, or other compact and stable formations. Thus, it is possible to imagine the flow of life in two dimensions. The first is a form hidden from the eye and is termed genotypic life. This is the dynamic existence of virtually unchanging genomes of all species of living organisms. The second is visible phenotypic life, which is the periodic phenotypic manifestation of genomic activity as various living bodies. One can say that various forms of cells and living bodies serve as distinctive *phenotypic frameworks of genomes*.

Phenotypic life is very changeable and is in a state of constant development. An infinite number of species have changed on the Earth over hundreds of millions of years. At one time fish 'reigned', then came dinosaurs, large predatory mammals, and now humans. In principle, these are all links in one chain of animal evolution. Although the phenotypes of given organisms differ significantly, their genotypes have gone through only modest molecular changes. The qualitative and quantitative composition of the nucleotides, as well as the chromosomal composition of karyotypes, has changed only slightly. In other words, as stated by Timofeev-Resovski, relatively small evolutionary changes in the genetic apparatus are greatly amplified in the process of expression of hereditary information, and manifest themselves in significant modifications in phenotype.

Thus, phenotypic life on our planet is a very old, widespread, diverse, stable, and changeable phenomenon. The reason for the diversity of millions of organisms (phenomes) is the presence of a similarly large number of variations in their genomes. On the basis of common origins and the unified principle of the organization of hereditary mechanisms, it is possible to say that all genomes of a given species and all genomes of all species in general are integrated into the system of the *Global Genome* (GG). This system consolidates discrete genomes of all living organisms into a complete entity. Analogously, the totality of all the phenotypes of all the living beings inhabiting the Earth can be imagined as an integrated system of the *Global Phenome* (GP).

The essence of life is therefore a process of continuous existence, with the development of very stable dynamic complexes of diverse genomes. These genomes contain specific programs for the managed ordering of matter, and the creation of specific incarnations of existence in the form of concrete representatives of distinct types of living bodies. From this point of view, the phenomenon of life can be defined as *a process of the continuous existence of evolving DNA (and in some cases, RNA) with various forms of its phenotypic (bodily) manifestations.*

## 1.2 Phenotypic Diversity and Genotypic Unity

There are currently several million known species of living organisms that inhabit the Earth, and more and more new species are being discovered each year. These organisms are very diverse in terms of their organization, function, metabolism, motion, habitat, and so on. They range from extremely small bacteria and unicellular organisms, to a multitude of highly organized multicellular plants and fungi, to millions of species of animals. Organisms also differ in terms of the units from which they are formed: the cells of plants, fungi, animals, and bacteria have significant diversity in their structure, properties, and functions as well. Fundamental differences between organisms are linked with peculiarities in the organization of their genomes.

The 'non cellular' part of the organic world is represented by a large variety of *viruses*. Viruses can be found in all kinds of shapes: rod-shaped, spherical, oval, etc. Viruses are one of the most widespread forms of existence of organic matter in the world by number. For example, the water in the oceans contains a huge number of bacteriophages—about 250 million particles per milliliter of water, and there are hundreds of thousands of species of viruses, most of which have not been studied.

Viruses are extremely small (15–300 nm) non-cellular parasites that consist of a single molecule of either DNA or RNA, several enzyme molecules, and a capsid. Different virus species vary significantly in their structure and organization, the DNA or RNA content, the host and place of parasitism, and the mechanisms of reproduction and proliferation. Moreover, they do not exhibit any metabolic activities. Some viruses provoke diseases in animals, while others act in fungi, plants, and even bacteria (bacteriophage viruses). Although they can be found everywhere, even outside of the cell, viruses do not display the properties of living things. However, they are capable of mutating, and therefore are also capable of adaptation and evolution. Viruses are the derivatives of cells, the universal carriers of genetic information within the world genome, and they are the representatives of genotypic life.

The most ancient cellular beings that still inhabit our planet are the prokaryotic organisms of *archaea* or *archaebacteria*. They do not have a defined nucleus or membrane-bound organelles, which is why they are considered to be closely related to bacteria. Today, it has been proven that these creatures are equally distant from both pro- and eukaryotes, and they are a unique fragment of the relict micro-world that populated the Earth 2–3 billion years ago.

Although the membranes of archaea are formed by phospholipids, as they are in bacteria and eukaryotes, there are several distinct differences in their composition and properties. Many archaea are capable of photosynthesis, but they lack chlorophyll. Instead, bacteriorhodopsin serves as the photosynthetic pigment. Only archaea are capable of photoheterotrophy, i.e., the use of solar energy for catabolism of foreign organic matter. Unlike bacteria, the genome of archaea contains introns, and this is one piece of evidence that indicates that eukaryotes originated from archaea rather than bacteria. In addition, their ribosomes are similar in size to the ribosomes of both eubacteria and eukaryotes. It is also characteristic for archaea to lack electron-transport chains, and the proton gradient is generated with the help of the bacteriorhodopsin proton pump. A unique peculiarity of some archaea is the possession of a complex of enzymes that are able to carry out methane genesis, something neither eukaryotes nor bacteria are capable of.

The majority of archaea are extremophiles: they have preserved their adaptation to the same conditions that were present on the Earth billions of years ago. Many hot springs, for example, are known to contain archaic thermophiles which are resistant to temperatures from +45 to +113 °C; archaea psychrophiles are capable of propagating at relatively low temperatures (from −10 to +15 °C); archaea acidophiles live in acidic environments (pH 1–4), whereas alkaliphiles prefer basic (alkaline) conditions (pH 9–11). Barophile archaea can even survive pressures of up to 700 atmospheres, and galophilus lives in saline solutions with NaCl contents of up to 20–30 %.

The group of life-forms known as *bacteria* includes a great diversity of species of free-living, simply organized, single-celled organisms (bacteria and cyanobacteria) which also tend to lack a well-defined nucleus and membrane-bound organelles. Bacterial genetic material is generally found as single linear or circular molecules of DNA that freely traverse the cytoplasm, as well as in various plasmids. Cytoplasmic organelles include mesosomes, thylakoids, and other diverse vesicles. These organisms have diminutive sizes ranging from 0.3 to 30 μm, and can be found in various shapes such as spheres (cocci), eiloids (*Treponema pallidum*), rods (tubercular bacillus), and so on. Some bacteria are surrounded by a dense capsule and may contain cilia or flagella for locomotion. Bacteria have also mastered multiple forms of energy conversion and synthesis. They are capable of both oxygenic and anoxygenic photosynthesis (i.e., where no oxygen is produced during photosynthesis), synthesis of organic matter from non-organic sources, and use of energy from mineral oxidation (nitrogen, sulfur, iron, or manganese), and they can even act as heterotrophs, although without phagocytosis. Representatives of bacteria are widespread on our planet. As with viruses, they permeate the entire biosphere, from miles down underground to the uppermost layers of the atmosphere. Some bacterial species are pathogenic and can therefore cause infections and diseases.

Prokaryotes, along with protozoa, make up 50 % of the planet's biomass. The geophysical composition of the modern biosphere is primarily attributable to the bacteria and archaea which have been acting on it for over 3 billion years. Eukaryotes and multicellular organisms simply claimed the prokaryotic biosphere as a habitat during all subsequent steps of evolution. Even today, the rest of the living world could not exist without microorganisms, as they remain the foundation of the planet's life maintenance system.

*Protista* and *protoctista* comprise the largest and most diversified portion of the eukaryotic world. This group contains several hundred genealogical branches of the tree of life that have never quite become true metazoans, with some rare exceptions. Because various protists are not strongly related to one another, phylogenetic taxonomy does not consider them to be a single group, separating them instead into many sub-kingdoms. The unifying feature of protists is that, whether they are unicellular or multicellular, they have a very simple organizational state which differentiates them from other eukaryotes.

The sizes of protista usually range from several dozen micrometers to several millimetres, although some 'achievers' can be gigantic in size. For example, komokinea can be several centimetres long, slime molds can range from 1–5 m in length, and brown algae can grow to as much as 35–70 m. Despite their relatively low level of organizational complexity, protists tend to have a very rich set of intracellular organelles, thousands of different enzymes, and a rather complex metabolism. Many organelles have special locations and characteristics. These include ejectosomes, pyrenoids, spiracles, axostyles, parabasal bodies, etc. Some Protista (foraminifers, diatoms) are encased in an external skeleton, or, on the contrary, have an axial endoskeleton (radiolarians and dictyochales). The composition of such skeletons can also be highly variable, containing such things as silicates, calcium saline, magnesium saline, and even strontium saline. Protista are found in a very wide variety of environments including aqueous media, soil, as well as in the other organisms. However, because the majority of them require water in some form, only a few are capable of living directly on the land or in the air.

*Fungi* or *mycota* are a large group of eukaryotes that currently includes about 100,000 species, although it is assumed that there are at least 10–15 times more. These are relatively simply built unicellular and multicellular organisms which differ in various osmoheterotrophic ways of feeding: they absorb nutritious material through their surface (like plants), but they are not capable of synthesizing organic matter from non-organic sources (like animals).

Cells of true fungi possess the majority of traditional eukaryotic organelles (except for plastids), as well as a number of specific ones. Fungi have cell walls which are composed of chitin and $\beta$-1,3-glycan. They also utilize the polysaccharide glycogen as a reserve energy source, the metabolic product of which is urea. Fungi have an excellent enzymatic composition, allowing them to dissolve lignin, keratin, and cellulose. They can also absorb and dissolve glass, rubber, and plastic. The scope of the fungal genome is the smallest among eukaryotes. Fungi can be found living in all kinds of environments and can be seen deep in the sea, in the soil, in the atmosphere, and even in many animals, plants, and other fungi.

The body of most fungi is composed of mycelium that consists of thin branching filaments called hyphae. The mycelium of pileate fungi is located in the soil, and forms a biomass on the surface known as the carposome. Fungi multiply asexually with mycelia and sexually with spores, and can be segregated into higher and lower distinctions. The hyphae of lower fungi, for example, do not have a multicellular structure (coenocytic), but are instead giant and intensively branched single cells with numerous nuclei (e.g., mucor, which forms mold on spoiled products). The hyphae of higher fungi, on the other hand, have a multicellular structure (septate), which consists of very long cells with multiple nuclei and a set of organelles typical of eukaryotes. The genetic material of fungi is also more complex than that of protista.

*Plantae* comprise hundreds of thousands of species of colonial, unicellular, and multicellular organisms. They are capable of oxygenic photosynthesis, during which they absorb photons of light and transform their energy into the energy stored in the chemical bonds of ATP molecules. These molecules are further used for the synthesis of primary organic molecules along with $CO_2$ and $H_2O$ from the environment. However, it is important to note that, at the present time, only organisms that contain chlorophyll *b* are considered unequivocally to be plants. These include green algae and vascular plants. The rest of the photosynthesizing eukaryotes are a part of the diverse world of the protists. True plant cells also characteristically have a double membrane (while photosynthesizing protista can have 1, 3, or 4 membranes) and a cellular wall composed of cellulose (in protista it can be lacking, or can incorporate $\beta$ -1,3-glycan, $\beta$ -1,4-glycan, proteins, and minerals). Furthermore, vascular plants do not have centrioles, although green algae do.

Plants have the largest biosynthetic potential in the organic world. They are found both in water and on land, and have the ability to regulate the water content in their bodies (poikilohydry). Plants also possess a rich system of symbiosis with other organisms such as bacteria (nodule diazotrophy), fungi (lichen and mycorhiza), and animals (pollination and dissemination).

Plants can also be further differentiated by their levels of complexity. Higher order plants possess complex organ/tissue structures, while inferior plants, represented by various algae and lichens, do not have tissues or organs, and usually exist as single cells, a cell colony, or a frond.

*Animalia* is the largest group of eukaryotes, and is represented by 1.5 million different species of multicellular organisms, as well as by a small group of their unicellular ancestors, the choanoflagellates. Animals are mostly phago-heterotrophic, which means that they are capable of absorbing organic matter from the environment through ingestion. Most also possess the ability to move autonomously, have a well-defined growth capacity, and produce urea, ammonia, or uric acid as the final product of protein metabolism.

Animal cells have a limited set of specialized organelles. They do not have rigid walls, which allows them to be elastic and movable, and they do not contain plastids or mineral inclusions. Most animals possess highly complex and specialized organ/tissue systems such as the nervous, endocrine, and immune systems, as well as others. They also have complex behavioural reactions, which are structurally conditioned due to the presence of neural networks.

Animals possess a tremendous variety of different types of cells that participate and specialize in distinct functions. Compared to the other life forms previously mentioned, animal cells possess an extremely complex and enormous genetic apparatus with many chromosomes. The various types of animalia are categorized on the basis of their structural and vital peculiarities (e.g., coelenterates, worms, mollusks, arthropoda, chordates), and every type of animal (e.g., mammals) has many different species within the greater classifications.

It should be noted that during the 3.5 billion year period of life on Earth, many millions of other species of living beings have inhabited our planet, as attested by their numerous fossil remains. All these organisms had the same 'nucleic-protein' foundations of organization as modern organisms, and every currently living organism appears to be a derivative of those from the past. Each one appears to be a link in a never-ending branching chain that stretches back over billions of years to the moment when life's processes first became associated with organized material systems. This chain continues to develop into an unpredictable and

**Fig. 1.2** The life star. A
cellular concept of life's
organization provides the
foundation for the five
kingdoms of living organisms

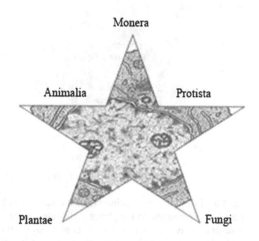

Monera

Animalia                    Protista

Plantae                     Fungi

impossible to imagine future. Reproduction of the first cells and their derivatives
formed the continuous chain of living organisms, capable of passing on their
structures and processes of life to the next generation. Colonies of new cells were
formed from certain types of particular cells. These new colonies possessed
properties distinctive to animals or plants. After millions of years, such colonies
originated the great number of living organisms with properties that arose from
their adaptation to environmental conditions. As a result of these processes, various organisms emerged on our planet wherever conditions provided aqueous
solutions of NA and proteins.

Despite their tremendous diversity, all living organisms have possessed (and
still possess) common general characteristics of organization and function. In
particular, all organisms are composed of cells (see Fig. 1.2). Every cell has a
membrane, cytoplasm, genome, a common mechanism of genetic expression,
similar rules governing inheritance and mutability, and analogous biochemical
processes. All cells transform energy, synthesize comparable compounds, and
maintain homeostasis. Cells of all organisms have similar molecular compositions
and metabolic mechanisms. Therefore, life appears to be a totality of structures
and processes that are identical in essence for all living bodies. Although the
structures and mechanisms may vary significantly depending on the species of the
organism, they do so within the confined parameters necessary for life. Basic unity
within the diversity of organisms indicates the identity of life in all its manifestations, a common source of origin, and its consecutive complications according to
the process of progressive evolution.

However, the major similarity between all living bodies is that every organism
appears to be the unity of genotype and phenotype, or, better to say *genome* and
*phenome*. In other words, a living body (phenome) is the final 'product' of the
realization of the mechanisms of a genetic program. The great diversity of DNA
and RNA molecules allows for infinite amounts of genetic information. Functioning separately or in various combinations, these molecules form the vast
number of genomic options. By means of protein molecules, genomes organize the

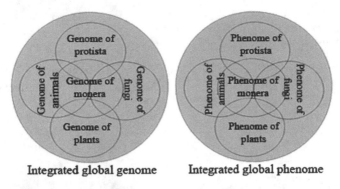

**Fig. 1.3** Two integral constituents of the integrated life system. The totality of all the genomes of all living creatures on the planet constitutes an integrated system of the global genome. It can be displayed as a nucleic net (genet) that covers the whole planet. Everything is interrelated within this net, and modifications in any part of it may cause unpredictable fluctuations in many other parts. The totality of phenomes of all the living creatures constitutes an integrated system of the global phenome, represented by the integrated protein net (phenet)

surrounding material space, creating specific phenomes that establish the individual representatives of a species of living organisms.

Thus, it is convenient to notice again that the totality of all the genomes of the different species on Earth can be interpreted as the integrated system of the Global Genome (Fig. 1.3). The expression of its separate discrete parts (genomes of individuals) leads to the specific manifestation of living bodies—phenomes. Therefore, the totality of all organisms of living nature presents itself as the integrated system of the Global Phenome. Thus, 'all life on Earth' appears as the global phenome which, in turn, is the product of the expression of the global genome. This statement indicates the integrity and unity of the phenomenon of life, despite the diversity of the individual discrete forms of its genotypic existence and phenotypic manifestations.

## 1.3 Expediency, Universality, and Similarity

The expediency of structural and functional organizations of living organisms is rooted in the correspondence between metabolism, physiology, morphology, behaviour, environmental factors, and the interaction between and interdependence of the environment and all kinds of different species. Besides that, every single organism shows an amazing logic in the organization, structure, and functions of all organs, tissues, and body parts, and also an amazing correspondence of the inner and outer structures with the external natural environment. For example, the constitution of fish is expedient exactly for the aqueous medium in which they live. These organisms have gills, fins, and streamlined bodies. Likewise, flying birds have adapted wings, fast responses, acute eyesight, light bones, and so on. Such apt correspondences represent the complementarity between form and function, between planet and life.

The successful combination of molecules, cells, tissues, organs, organizational patterns, and biochemical and physiological processes that came into existence millions of years ago after sorting through natural selection is characteristic of the majority of modern living organisms with different levels of organization. Such universality is typical for various biological structures, functional complexes, and biochemical and physiological mechanisms. The similar organization of living bodies is determined by the natural selection of the most auspicious combinations of molecules and molecular complexes, the most thermodynamically efficient and economical biochemical processes, structural shapes, concepts of interaction with the environment, and so on.

The significant influence of the range of constant environmental factors (such as gravity, photoperiodism, electromagnetic radiation, media, temperature, etc.) has come to be an important element of selection of universal principles of organization in biological systems. Given the above factors, nature uses optimal algorithms of organization based on the similarity principle. The evidence for this notion is the existence of a large scale of semblance and universality at all levels of the evolutionary hierarchy of organisms. It is apparent in similar plans of construction, metabolic processes, functions, and similar mechanisms for maintaining homeostasis.

The phenomenon of similarity in evolutionary biology is reflected in the definitions of homology and analogy of structures. Structures that have a genealogically common basis but are capable of performing various functions are called homologous. In contrast, analogous structures have a different basis and their similarity is determined by solidarity of performed functions. In this context it should be noted that the genomes of all living beings are homologous to each other, since, despite the diversity of performed functions and individual distinctions, they all have a common basis. At the same time, the phenomenon of analogy is widespread in the world of phenomes, which is prone to environmental pressures.

In particular, homology is found in the fundamental processes of life where nature uses the same limited standard set of molecules to build all life forms. The successful composition of substances composed of organic monomers and the formation of multifunctional polymers (nucleic acids and proteins) is typical for all living organisms. The presence of unified membranes, the principle of compartmentalization of living bodies, and the principle of autonomy are inherent for single cells as well as for multicellular organisms. The emergence of the cell as a result of evolution represents the universal standard unit of organization of all living beings.

Enzymes are the major characters in all micro and large scale (macro) processes. The identity of many molecular mechanisms and functions has remained without significant changes at all levels of organization in living systems from single-celled algae to modern mammals. As an example of the universality of life's mechanisms, controlled cellular events such as ATP formation and the transfer of electrons in various organisms rely on principles of selective catalysis, which are operated by virtually identical membrane proteins and enzymes.

Genetic likeness is determined by the presence of genomes as keepers and operators of genetic information, and by the presence of unified mechanisms of realization of this information. Without exception, the genetic material of all organisms is presented by nucleic acids that have unified concepts of organization, properties, and functions. Nucleic acids differ from one organism to another only by the sequence of the standard nucleotides and their relative abundance. A gene arises when the information stored in genomic sequences leads to the production of some other nucleic acid (such as mRNA) or protein product. The stipulation of a phenotype arising from the genotype is a key principle for all living things.

Living nature uses a quite limited number of metabolic pathways. The majority of them are conventional for most organisms (e.g., glycolysis, tricarbonic acid cycle, protein synthesis, etc.). Similar functions and roles are maintained by hundreds of similar enzymes. Nature also uses a limited amount of standard regulatory molecules (e.g., the somatotropic hormone has an identical molecular structure and mechanism of action for all mammals). All organisms are self-regulating systems; they are built and function according to the information found in their makeup and surroundings, which they can perceive, process, and exchange with the environment and other systems. These and many other universal principles of the way living matter is organized were chosen by nature billions of years ago and are typical for most modern organisms.

The expedience of the semblance of properties of different living bodies is determined by the necessity for every one of them to achieve the same strategic life tasks: survival, reproduction, and distribution. Successful versions of organization were fixed in various genomes, and due to the unity between the global genome and DNA continuity, these versions spread to all kingdoms of living organisms. These facts testify to the common nature and close relation of all living things and to the common origins of life. Life, therefore, exists as an integrated global network with molecules of nucleic acids as the connecting links and threads.

It is amazing how all living beings are perfectly complete organisms. For example, whether looking at single cells, worms, amphibians, or mammals, all of them, regardless of their level of organization, are rather rationally structured and excellently adapted in spite of the difference in quality and quantity of genetic material. This difference only affects the complexity of bodily structures, but does not affect the ability to adapt, survive, and multiply. Absolutely all the functions and metabolic processes necessary for survival in certain environmental conditions are available at all levels. Basically, every genome is sufficient for the living body that it creates, and this body, in turn, is perfect to sustain the life of its owner.

# 1.4 Probability and Life

Cells are very complex, highly organized molecular systems that have strictly defined processes occurring within them. These processes take place in specific locations and directions. The probability of the emergence and existence of such systematic processes is extremely low, because the appearance of defined molecules in constrained qualitative and quantitative ratios in an accurate hierarchic order coupled with a proper location in a microscopic space that is separated from the environment is highly improbable.

However, improbable does not mean impossible or motiveless. Genetic and other informational programs considerably raise the probability of such material events. Because of this, low-probability events do not just happen once in a while, but steadily reproduce over and over again in living systems. In particular, specific structural and functional proteins and enzymes are synthesized according to specified genetic information. Structural proteins and other molecules become organized into certain probabilistic cellular macrostructures (e.g., organelles) according to the laws of chemistry and physics. It is these functional proteins and enzymes that direct biochemical processes from millions of possible directions to just those that are necessary for the cell to exist.

It is important to realize that the incredible complexity of multicellular bodies does not happen miraculously, but is formed during the developmental process. These 'incredible constructions' are built step by step from the fertilized ovum under the laws of development, based on the genetic programs and molecular mechanisms necessary for their realization. It is the mechanism of development that allows the realization of low-probability events that determine the existence and appearance of extremely complex living bodies.

Another reason and condition for the implementation of low-probability events in living organisms is the purposeful utilization of energy. The fact of the matter is that processes of destruction are more probable according to the laws of thermodynamics, because they do not require energy consumption. But any processes that serve to order matter for creation require certain forces and energy applications. Therefore, definite material structures and processes require an energy supply for the creation and maintenance of ordered biological systems. The utilization and employment of the energy is achieved by purposeful enzymes that target only the necessary biochemical reactions with energy. Therefore, an organized biological system can provide the conditions required for low-probability events to occur by exploiting effective and selective applications of energy.

The probability of specific chemical reactions in cells is significantly raised by the genetic selection of enzymes. These molecular machines do several important things: they catalyse strictly defined processes, elevating their probability several million fold, as well as increasing the speed of otherwise improbable processes many thousands of times. In this case the 'all or nothing' law is followed: where the enzyme is present, a well expressed random process is observed, while no enzyme means no process.

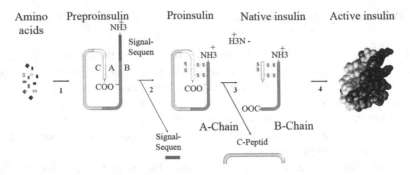

**Fig. 1.4** Main steps in the formation of active insulin molecules. (*1*) Polypeptide chain formation from 51 amino acids, (*2*) Signal peptide is cut from synthesized polypeptide preproinsulin, and the molecule is folded by two disulphide bonds, (*3*) Another peptide is cut from proinsulin, (*4*) Subsequently, polypeptide chains are spatially folded. The result is the active form of insulin

Various DNA mutations and recombinations that constantly occur in cells are examples of random events. Mutations can be caused by various factors that unpredictably influence any of the trillions of nucleotide pairs. The results of such mutations are also barely predictable and may be neutral, lethal, positive, or negative. Since mutations are the source of new characteristics for natural selection, it is clear that the direction of evolutionary processes is totally random and unpredictable, and would most likely correspond with certain environmental conditions.

We have already mentioned that the probability of formation of concrete macromolecules within cells is extremely low. Let us have a look at the formation of one of the smallest proteins—insulin (Fig. 1.4). It consists of 51 amino acids, so the probability of their conjunction is only possible on the order of one in $20^{51}$. And this is just one option of the $2.6 \times 10^{66}$ possibilities! Moreover, in order to be activated, insulin must obtain a well-defined structure with a spatial form during the process of post-translational modification. In order to achieve the active form of insulin, the initial polypeptide has to be cut by enzymes into several distinct parts, two of which are afterwards joined in a specified location by disulfide bonds. Only then does insulin assume its specific, biologically active form. The probability of obtaining such a unique variation is extremely low—just one in a trillion. Therefore, the probability of existence of every one of the tens of thousands of proteins in living organisms is close to zero. However, these proteins, even those that consist of hundreds of amino acids, exist happily and plentifully. Moreover, they are constantly reproduced with great speed, and are successfully inherited by new generations over millions of years.

The individuals of various species are also matters of chance. For example, let us evaluate the probability of the birth of a human individual. To do so, we will take into account only the following conditions[1]:

---

[1] *Numbers are rather approximate, but reflect the essence.*

The chance of meeting of the partners. The probability of a relationship between a specific man and woman in a two-million person megalopolis (if the sexual ratio is equal) is one in $10^{12}$ possible outcomes.

The chance of insemination of the egg by the spermatozoa. Women have in their ovaries tens of millions of allele-varying egg cells, every one of which can mature and be fertilized. Every man produces $10^{10}$ variations of allele-varying spermatozoa during the 50 years of his reproductive activity. Therefore, the probability of fertilization from two individual gametes is also a one in $10^{17}$ chance.

The chance of crossover events during meiotic formation of gametes. Human haploid chromosomes consist of approximately $3.2 \times 10^9$ nucleotide pairs. Therefore, the possibility of their recombination during contact with homologous chromosomes (taking into account only a single crossover) corresponds to a chance of one in $10^{19}$.

The chance of chromosome combination and disjunction into separate gametes during meiosis. The number of possible combinations can be as high as $10^5$.

Therefore, taking only these circumstances into account, the probability of the birth of Gennadiy Zhegunov is one in $10^{43}$. This makes such an event highly improbable (the total number of human individuals that have inhabited the Earth so far is approximately $10^{10}$). Meanwhile, in reality, each living person does in fact exist, despite the almost absolute improbability of this happening. These probabilities are similar for all other living organisms with sexual reproduction. Individuals, therefore, seem to appear by chance, unscheduled, without permission, and against their will.

Nevertheless, unlikely events in biology are quite substantial, having their own reasons, conditions, and circumstances. It is absolutely clear that random events can occur under certain conditions, but based without exception on the laws of nature. Furthermore, the probability of the implementation of the random processes of self-organization and evolution rises with time, so that randomness may indeed be realized over some prolonged period.

In summary, it can be stated that cells and organisms are biological systems that develop conditions for transformations of random events into ordered events—acting as *extraordinary enhancers of probability and chance*.

## 1.5  Temperature and Life

Thermal motion, as one of the forms of existence and transformation of matter and energy, is of particular importance for living bodies. And the temperature is a quantitative measure of this kind of motion, which determines the boundaries of life.

The temperature range of the universe is extremely wide. It varies from absolute zero ($-273.15$ °C) to many millions of degrees Celsius. On Earth, such limits are much narrower, ranging from $-88$ in the Antarctic to 5,000 °C in the

Earth's core. The boundaries of the temperature range of the oceans, seas and rivers, and groundwater are considerably narrower, and range from 0 to 100 °C. This is the temperature range where all the living bodies exist.

Single-celled creatures, primitive multicellulars, fungi, plants, and poikilotherms cannot maintain a constant body temperature. However, they are still capable of possessing viability and activity within the aforementioned temperature limits, since their enzymes operate stably in the specified range. Birds and mammals are able to maintain a constant body temperature automatically in a very narrow range from 32 to 40 °C. Under such conditions, biochemical and biophysical processes in living systems possess stable behavior regardless of environmental conditions, which gives these classes of animals significant advantages.

Temperature plays a significant role both for the existence of living bodies, and for the existence of the phenomena of life. Both occur in the same temperature range from 0 to 100 °C, which is the range of liquid water. The substrates of life, the NAs and proteins, may also exist for a long period in a frozen state at temperatures down to absolute zero. Under such conditions they do not manifest their biological properties and do not possess the phenomena of life, but retain the structure and ability to realize their life potential. In this form the "seeds of life" could travel in outer space on the fragments of planets for millions of years. But at temperatures above 100 °C, there is no phenomenon of life, because the existence of living bodies becomes impossible. Most of the macromolecules of cells become destroyed under such conditions. And what is most important, there is a phase transition of water into a gaseous state, which is not propitious for life.

Thus, it should be emphasized that most of life is realized within the narrow confines of temperatures which are close to the lower boundaries possible in nature (0–40 °C). At such low temperatures, many biochemical reactions are thermodynamically impossible or can only occur at very low rates which cannot provide for the manifestations of life. Only the presence of enzymes, which accelerate reactions a thousand fold, can enable the implementation of the necessary biochemical processes. Only with the help of biological catalysts is the world of biochemical reactions determined, leading to the manifestation of life. By means of high rates and specificity of action, enzymes distinguish only a limited set from an unlimited number of possible reactions between the countless molecules. Thus, the enzymes increase the probability of the processes, which are otherwise unlikely within the vital temperature range, thereby providing for the incredible phenomenon of life.

The upshot of this is that low temperatures constitute a very important condition for the emergence and manifestation of life. Only against the background of a "slowed-down" environment, where the transformation of matter and energy flow sluggishly, can specifically made and naturally embedded enzyme molecules clearly distinguish a limited number of interconnected chemical reactions. Chemical reactions then become biochemical and life is generated from the non-living.

# Chapter 2
# Material Basis of Life

## 2.1 Material Nature

Living beings, as a part of nature, are derivatives of the developing material world. All biological processes, including those that constitute the existence of living beings, occur within the limits imposed by natural laws.

Virtually all the elements of the periodic table have been found in various representatives of living organisms. The main elements of living organisms are: carbon, hydrogen, oxygen, nitrogen, sulphur, and phosphorus. These elements are widely spread around our planet and it is exactly these elements that form organic compounds and basic macromolecules: proteins, nucleic acids, carbohydrates, and lipids. Moreover, 70–80 % of the mass of most living bodies consists of water and various mineral and organic salts. Thus, all living organisms consist of the same elements and organic and inorganic molecules as non-living bodies.

The properties of various separate molecules that compose cells and organisms do not differ from the properties of the same molecules in non-living systems. However, specially arranged and organized complexes of macromolecules (such as cell compartments) possess completely new 'biological' properties. As a result, biological objects differ significantly from non-living bodies, having unique properties that are only typical for living beings. For example, they are able to reproduce, take in nutrients, respire, and so on. These functions, as a matter of fact, are stipulated by the arranged interaction of molecules and cells that compose an organism.

It is known that organisms possess a set of physicochemical features. In particular, they have a discreteness and hierarchy of structure, and an interaction and interdependency of parts which establish the concept of integrity. An organism's structure has a molecular basis, and it converts and uses energy for accomplishing work, etc. Due to the physicochemical foundations of life, the study of biological objects can be accomplished with the application of powerful modern physical and chemical methods. This has led to the discovery of many molecular and cytogenetic mechanisms of the phenomenon of life.

G. Zhegunov, *The Dual Nature of Life*, The Frontiers Collection,
DOI: 10.1007/978-3-642-30394-4_2, © Springer-Verlag Berlin Heidelberg 2012

Thousands of biochemical reactions occur in all organisms on the basis of chemical laws (e.g., the law of conservation of mass and energy). The majority of substances in an organism are dissolved in water, and the mechanisms and behaviours of dissolved substances are the same in a cell or in a test tube. The presence of enzymes is necessary for almost all biochemical processes. Enzymes act as chemical catalysts. Various factors, such as light, temperature, pressure, etc., affect biochemical reactions the same way they affect chemical reactions outside biological systems.

Hormones and neurotransmitters are chemical molecules of a certain nature and structure. They transmit definite signals and information by associating with receptor molecules on the surface of cells and changing physical states of a membrane and/or other molecular complexes. Through this process, the cells of multicellular organisms communicate using a physicochemical language.

The chemical interactions of various molecules are the foundation of life. Features and properties of living organisms, recorded in their DNA molecules, are stored and transmitted by chemical means. Biochemical transformation (that is, the transformation of certain molecules leading to the formation of others) appears to be the essence of the majority of biological processes, these being founded on the mechanisms of chemical bond breakage and formation. These mechanisms are linked with the interaction and exchange of elementary particles (protons and electrons) and atoms between the reacting molecules.

Motions and interactions of molecules in living systems involve physical processes such as diffusion and osmosis. These processes are the result of thermal motion. The basis for all types of movement in living bodies is the thermal motion of molecules that appeared along with matter during the Big Bang. Living organisms, like all other material bodies, exist in a certain time and space. They are complex systems that exist due to the global properties of all their constituents in motion.

Molecules and supramolecular cellular structures also possess various functionally significant physical properties (e.g., polarity and hydrophobicity). Cellular membranes have electric potentials, as exemplified by neural cells which function on electrical impulses. Blood movement occurs according to the laws of rheology. Joints, skeletal bones, and muscles act on the basis of mechanics. Energy transformation occurs according to the laws of thermodynamics. All of the processes and elements of an organism such as vision, hearing, nerve impulse conduction, permeability of different substances, and so on, are dictated by physical laws.

It is important to stress that all living bodies possess physicochemical characteristics and that all the processes in living organisms obey the laws of physics and chemistry. To put it another way, the laws of physics and chemistry are not violated in biological objects.

The major difference between the scientific and non-scientific (religion, mysticism, etc.) points of view is that scientists do not consider life to be a separate (or different) manifestation from the rest of the material world. They consider it to be a particularly organized part of nature that has the same material basis as the rest of the world. Therefore, a strong integration occurs between the biological and physical worlds. Living matter is a part of nature that follows the same universal physical laws as non-living matter. These laws, in turn, determine certain limits for the properties and features of living beings.

## 2.2 Energetic Basis (A More Detailed Description of this Topic will be Given in Part 3)

Life is hard work. It is a constant battle fought by organized living systems against the destructive actions of unfavourable factors from external and internal environments, and against the natural expansion of entropy. It is also a task involving permanent massive synthesis of various organic substances needed to maintain structural and metabolic homeostasis. A colossal amount of work also takes place to support the various functions of an organism: motion, respiration, nourishment, reproduction, growth, etc. Therefore, the physical concept of 'work' is applicable to every process in living systems, whether it is at the molecular level or at the level of a whole organism. It is obvious that the implementation of any type of work requires energy, and for the implementation of such a multifarious amount work, a lot of energy is needed.

To this end, all organisms constantly transform, accumulate, and purposefully use free energy obtained from the environment. The inflow of energy gives cells and organisms the opportunity to function and also to support a high level of order and homeostasis for long periods. If the energy intake into the organism decreases or stops, its structure gradually deteriorates, functions become violated, and the organism eventually dies. Liquid, flexible, and vulnerable bodies that live in unfavourable surroundings manage to maintain their organization and properties for many years only by means of constant work against natural forces of destruction.

## 2.3 Nucleic Basis (A More Detailed Description of this Topic Will be Given in Part IV)

Nucleic acids (NA) are unique molecules that are essential for every organism (from viruses to man) for storage, application, and transmission of genetic information. Nucleic acids are the basis of the structural and functional organization of genomes. They contain genes which determine the synthesis of proteins. These proteins, in turn, determine the character and peculiarity of metabolism, natural laws of growth and development, various functions, etc. The mechanism of NA duplication and transmission of copies to filial organisms lies at the heart of reproduction in the sense that the flows of genetic information determine prearranged transformations of organic substances and the self-organization of material

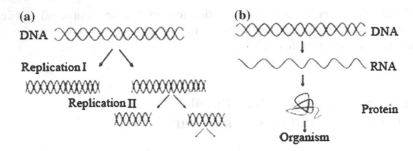

**Fig. 2.1** Nucleic acids, the stem molecules of life. (**a**) Molecules of DNA possess an unlimited ability to reproduce themselves. They form the stem of replication or reproduction (horizontal stem). (**b**) Molecules of DNA are the fathers of RNA and proteins, conditioning the appearance of all the necessary molecules in cells. They form the stem of expression or stem of production of living bodies (vertical stem)

space in biosystems. The slightest abnormalities in NA structure may lead to unfavourable consequences or even the death of the organism.

There are two main groups of nucleic acids: DNA and RNA. Both are biopolymers consisting of monomeric nucleotides. DNA and RNA differ in their chemical structures and biological properties. Genetic information is encoded by DNA molecules, and the RNA molecules act as intermediaries that translate the information for the synthesis of corresponding proteins. It has been shown that certain molecules of RNA do not participate directly in protein synthesis. Possessing catalytic abilities, these molecules take part in the regulation of transcription, splicing, translation, and modulation of protein functions and special conformations. These molecules are therefore capable not only of carrying the genetic information, but also catalysing chemical reactions and regulating many cellular processes. RNA seems likely to have been the organizational foundation of life, long before the emergence of DNA and other enzymes. RNA is the only substance known to possess properties of both an informational matrix and a catalyst. The discovery of catalysing RNAs by T. Check and C. Altman is one of the most significant achievements of biology to date.

The foundation of life lies in the abilities of nucleic acids to store and act on information. The main molecule of life, DNA, is the keeper and supplier of genetic information. It is thanks to DNA that living organisms are able to reproduce, grow, and function. By analogy with stem cells, DNA molecules may be called *stem molecules* (see Fig. 2.1), since they are the forefathers of all the trillions of other organic molecules from which every organism is built and through which it functions. RNA is the intermediary in the biological information pathway leading to the production of proteins. Cells, organs, and bodies are built from proteins possessing a wide variety of properties and functions. In this case, the flow of genetic information is directed from the past to the present, from encoded genetic blueprints to realized protein materials.

During reproduction, genetic information is duplicated by means of DNA replication, and then, as a result of cell division, it is passed to filial organisms. In this case, genetic information becomes hereditary. This information flow is directed from the present to the future, from parental DNA to filial DNA.

These two pathways for the flow of genetic information are crucial for the existence of living bodies. In their foundation lie structural and functional transformations of nucleic acids. Thus, nucleic acid molecules are the substrates of life, unifying material and informational bases for all living beings. The operation of nucleic acids appears to be the essence of life.

## 2.4  Protein Basis

Proteins are the foundation of a cell's structure and function. They constitute up to 20 % of the mass of living mammals. Almost all the remaining mass, approximately 70 %, is attributed to water. The final 10 % consists of other organic and inorganic substances including nucleic acids, lipids, and polysaccharides. Proteins define the structure and shape of a cell, determine the functions of organelles and even cells themselves, and act as tools for molecular recognition. Enzyme proteins, for example, participate directly in metabolic processes, energy exchange, and the maintenance of homeostasis.

There can be up to 50,000 different proteins in mammalian organisms. The specific presence of a given set of proteins in a given organism determines its specific features of structure and function. For instance, protein sets that determine morphological, metabolic, and functional features would differ in undifferentiated and differentiated cells, even from the same lineage. In other words, proteins are the primary phenotypic traits.

The totality of structural and functional proteins of a cell is defined as the proteome. It is a complete set of tens of thousands of different proteins that provide myriads of possible processes and functions. The qualitative and quantitative composition of the proteome is determined by the genome—the totality of genes of the given karyotype. The structure and properties of an organism are predetermined and dependent on protein quantity, composition, localization, peculiarities of their structures, and physical and chemical properties.

The major conceptual point here is that proteins are *a nexus between the genome and the phenome*, that is, between the virtual genetic program in the DNA molecules and the real material body created by this program: genome → protein → phenome. Proteins are the intermediaries in the realization of genetic information, the tool and the main element in the creation of living bodies—'a phenotypic framework of the DNA'. Proteins, in other words, are the structural and functional foundations of a phenome.

Life would be impossible without the properties of proteins. It is therefore considered that life on Earth has a protein nature. In this context, the definition of life as a mode of existence of proteinaceous matter, as stated over a 100 years ago, is quite warranted, although today we need to add the known role of nucleic acids and enzymes to this notion.

## 2.5 Aqueous Basis

Not a single organism is known which can or could have existed without water. Moreover, water is the main substance of most living bodies. As previously stated, water comprises up to 70 % of their total mass on average. The significance of this aqueous substance is determined by its unique physical and chemical properties.

Aqueous molecules associate with each other via hydrogen bonds. They form an integrated phase ('universal ether') in which biosystems exist and all of life's processes occur. Water is a very effective solvent, a property arising from the high polarity of its molecules. The majority of organic and inorganic molecules in cells occur in a dissolved ionized form. Even insoluble substances change to a colloidal or emulsified state, which allows them to interact with the aqueous medium. Such conditions significantly increase their flow dynamics and reaction capabilities, which are extremely important for ensuring proper metabolic and physiological processes. Water and its ionization products (such as $H^+$, $OH^-$, and $H_3O^+$) significantly influence the properties of many cellular components. In particular, they affect the structure of proteins, nucleic acids, enzyme functions, organization of membranes, etc. Aqueous media allow for high velocities to be achieved by molecules in Brownian motion, as required for molecular interactions. Water also participates in many biochemical processes as a substrate, and is generated as an end product in many biochemical reactions. For example, water participates in aqueous photolysis, without which the process of photosynthesis would be impossible, as would the emergence and existence of flora and fauna. Water also exhibits a capillary effect, i.e., fluidity in very thin channels. This is quite important, for example, in the exchange of matter between cells and blood capillaries. The heat capacity of water exceeds the heat capacity of any other biological substance, allowing it to act as a thermal balance regulator in organisms. Water also serves as an environment for millions of different species, and life substrates can only exist temporarily or in a diminished capacity without it.

It should be noted that there is no water in a free state (e.g., a glass of tap water) in living organisms. Most of the water present in cells is in a bound state. Due to the polarity of molecules (proteins, NA, carbohydrates, amino acids, anions, cations, etc.), they bind and orientate a substantial amount of water molecules in several layers at their surface. Therefore, the inner content of cells is represented by a colloidal solution (gel), which is mainly composed of proteins. Colloidal solutions differ significantly from true solutions. In particular, gels have an inner organization due to the ordered orientation of water molecules around proteins that form the intracellular skeleton. Because of this, cells gain dual properties of liquid and solid bodies: density and constancy of shape and structure, as well as high plasticity. It is due to these unique properties that closely connecting mammalian cells form large, almost rigid organisms, despite the 70 % aqueous content of each cell.

Bound water also has the ability to participate in chemical and physical processes, which allows for practically unimpeded migrations and interactions of

**Fig. 2.2** The triangle of life.
The basis of life lies in the
existence of aqueous
solutions of proteins and
nucleic acids. From the
aqueous point of view, living
organisms are highly
organized, long-living
colloidal solutions

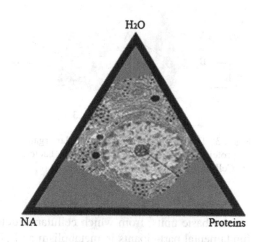

substrates and metabolites. Moreover, the intracellular organization of water has required the formation of special 'cytosolic channels' that participate in the purposeful displacement of certain hydrated molecules across the cell interior. The displacement of various molecules such as nucleotides, ions, amino acids, etc., is achieved through the use of these channels. There is also experimental evidence to suggest the presence of controlled local phase transitions (gel to liquid and vice versa) of certain cytoplasmic areas that are capable of directed substance transfer. Such properties of intracellular water establish internal mobility, which may be the basis of all processes and mechanisms in living bodies (see Fig. 2.2).

Therefore, the importance of intracellular water is highlighted by the unique duality of its existence as an ordered liquid crystal structure with the simultaneous retention of liquid properties.

Life has emerged, progressed, and exists in aquatic environments. Not a single manifestation of life is known to exist without water. Water thus plays a crucially important role in the establishment and perseverance of life. It is for this very reason that scientists search so carefully for $H_2O$ on other planets and satellites of the Solar System, because it appears to be the main criterion for the emergence of nucleic-protein life.

## 2.6  Cellular Basis

Cells are very complicated biological systems that possess a high degree of order and the capacity to develop, maintain integrity, and reproduce. Cells are mobile, open, non-equilibrated systems that represent the integration of structure and function. The content of cells is so dynamic that they can be considered rather as continuous processes than physical bodies. The basis of cellular life is grounded in the incredible orderliness of macromolecules which create complex supramolecular structures (membranes, organelles) possessing certain functions. Protein molecules

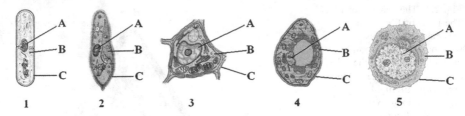

**Fig. 2.3** The common principle of cellular organization. (*A*) Genome. (*B*) Colloidal matrix based on proteins. (*C*) Membrane. (*1*) Cell of a bacteria. (*2*) Cell of a protozoa. (*3*) Cell of a plant. (*4*) Cell of a fungus. (*5*) Cell of an animal

are the basic units, from which cellular structures are built, and enzymes are the fundamental participants in metabolism and all its associated functions. Likewise, qualitative and quantitative cellular compositions are controlled and regulated by nucleic acid molecules.

Cells are the basic structures of unicellular organisms, prokaryotes, fungi, plants, and animals. Despite the differences in structure, metabolism, and function of cells from the various kingdoms, all of them share many common principles of structure and organization (see Fig. 2.3). For bacteria, protozoa, certain algae, and fungi, the definitions of a cell, organism, and body coincide, since they are unicellular creatures. However, vast species of other fungi, plants, and animals are multicellular, because they consist of thousands, millions, or even trillions of cells. Many of these organisms have tissues and organs which arise from collections of cells with similar functions and structures. The coordinated performance of trillions of cells and tissues in animals, for example, has led to the development of various vital bodily structures such as the neural networks of the brain and the endocrine and immune systems.

Despite being a vital component of a tissue or an organ, every cell, first and foremost, lives its own life. The organization and metabolism of a cell is primarily directed towards the maintenance of its own homeostasis. This requires the expenditure of the majority of the cell's energy reserves, leading to low energy utilization for the needs of an organ or the organism. It is therefore legitimate to consider multicellular organisms as a complex community of tiny living bodies that are united for the purpose of mutually beneficial coexistence. This is a qualitatively new concept of the organization of biosystems and cellular existence, where cells must act as part of a large specialized colony. The cells must work in accordance with this, following the interests of an organism and maintaining its integrity and functional diversity while balancing their own needs in the process. The entire organism and all of its interrelated components are regulated and coordinated by the genetic programs of its unified genome. In other words, a multicellular organism is the product of the selectively expressed genome of different cells which exchange matter, energy, and information with the external environment.

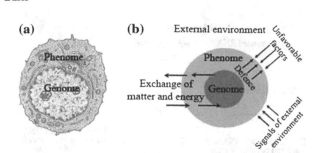

**Fig. 2.4** Schematic view of cellular organization from the standpoint of the genome in a phenomic frame. (**a**) Picture of a cell. (**b**) Interactions of the genome and its phenotypic frame with the external environment. The phenome is an active intermediary between the genome and the environment. Various cellular components accept signals from the environment and maintain the homeostasis of the genome. The phenome of multicellular organisms is composed of organized cellular masses and non-cellular substances. The external membrane of a cell body and the totality of metabolic and physiological processes of an organism withstand environmental factors, maintaining homeostasis of the body which incubates numerous genomes

Thus, every cell is an autonomous body that represents a dynamic solidarity of interacting macromolecules, organelles, and compartments, which form a highly organized and self-maintaining system on the basis of genomic programming.

The separate nucleic acids and proteins cannot alone set the conditions for life without a special medium where material transformation can occur. That is, separate molecules of nucleic acids and proteins cannot be considered alive unless they form a system that is integrated into the complex colloidal matrix of the protoplasm. The colloidal system of protoplasm, which consists of structured and compartmentalized elements, may be considered as a matrix. This matrix can be characterized as being 'cybernetic', meaning that it is defined as a complex of interconnected structure-informational elements (see Fig. 3.3). This matrix (nucleoplasm and cytoplasm) serves as an effective environment where genomic processes can occur to ensure accurate and directed movements of energy and matter and their controlled interactions for life. The functions and activities of the molecular matrix are directed by the genome via protein synthesis. The genome acts as a sort of biological microprocessor. It is therefore clear that the minimal components of living systems include not only nucleic acids and proteins, but also a complex aqueous-colloidal protoplasmic matrix.

Genetic information is an organizing and directing force which leads to high levels of order and functionality within a given cell. Genomes create around them an ordered environment called the phenome (see Fig. 2.4), which allows for the maintenance of genomic homeostasis necessary for the purposes of self-preservation, reproduction, and growth. A cell's composition is a reservoir for the genome, and has a set of properties and functions for the cell's successful existence and reproduction. Although the cell is an autonomously acting structure, its activities depend totally on genomic directions. The creation, appearance, and processes of new cells occur only on the basis of the genetic information stored

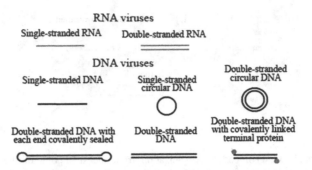

**Fig. 2.5** Several types of viral NA. Viruses may contain single- or double-stranded NA molecules. DNA molecules of certain viruses may also be annular. Prokaryotic as well as eukaryotic organisms may be hosts of all these types of viruses. Some examples are: single-stranded RNA—bacteriophage R17, poliovirus; double-stranded RNA—reovirus; single-stranded DNA—parvovirus; single-stranded circular DNA—M13 and X174 bacteriophages; double-stranded circular DNA—SV40 and polyomaviroses; double-stranded DNA—T4 bacteriophage, herpes virus; double-stranded DNA with covalently linked terminal proteins—adenovirus; double-stranded DNA with covalently sealed ends—poxvirus

within their genomes. The genome, therefore, creates the phenome as an intermediary between genotypic life and the external environment.

We can conclude that the *genome* is a conceptually defined specific portion of a cell that contains a select set of NA and proteins united into an integrated structural and functional system. This system contains special genetic information as well as mechanisms and instruments for its use. Analogously, the *phenome* is a conceptually defined portion of a cell that surrounds and integrates the genome, forming a monolithic body. For unicellular organisms, these are whole cells with the exception of a genome. For multicellular organisms, these are whole bodies, which are highly organized colonies of standard genomes in a phenotypic framework.

## 2.7  Non-Cellular Forms of Life?

Our planet is inhabited by millions of different species of living organisms. An overwhelming majority of these organisms have a cellular structure, but the rule does have exceptions. Some scientists believe that non-cellular forms of life are represented by viruses. This is a vast group of 'microscopic beings' that possess biological properties, but do not have a cellular structure. Viruses are tiny bodies composed of some sequence of nucleic acids (DNA or RNA) (Fig. 2.5), which is usually surrounded by a coat formed from several different protein molecules (capsid). Some complex viruses are additionally covered with a membrane derived from lipids, proteins, and carbohydrates (supercapsid). The genomes of the simplest viruses consist of only three genes: a gene for a protein that causes the rupture of a host cell's membrane, a gene for an enzyme that provides for the replication of the viral genome, and a gene for a capsule protein. The NA content of different viruses can differ greatly. RNA from *bacteriophage* Qβ that infects *Escherichia coli*, for example, contains only 3,500 nucleotides. Other larger viruses may contain up to 200–300 genes and up to several hundred thousand nucleotides.

Viruses are capable of existing outside of cells without manifesting any vital activity. In this condition they are passive and incapable of reproduction. However, when they penetrate a cell, they become very active, and acquire such properties as reproduction and development. During viral infection of a cell, the nucleic acid of a virus is transported through the cell membrane by various means, depending on the type of virus. For example, this can occur through the binding of the virus to membrane receptors on a host, followed by endocytosis (*Semliki forest virus*), or by means of injection (*bacteriophage lambda*). Nucleic acids of many viruses interact directly with the genome of a cell. Other viruses are replicated and translated in the cytoplasm (though indirect interaction with the genome of a host cannot be excluded). The genomic component of many viruses can integrate into the host cell's genome due to its intrinsic similarity to the host's NA constitution; it can be reproduced, transcribed, and translated using its endogenous cytogenetic machinery. New viruses are then assembled in the cytoplasm from the newly synthesized viral proteins and nucleic acids. Thus, viruses are 'alive' when they are inside a cell, but outside they do not display such properties, thereby demonstrating a form of intermittent life. In fact, it is more correct to describe a virus as a very active 'nucleic acid molecule' than as an actual living being inside the cell, since only the NA penetrates the cell, accompanied by a few proteins.

Viruses are, ultimately, intracellular molecular parasites that are capable of infecting nearly all representatives of living organisms. The internalized NA modifies the inner space of a host cell, making its genome and phenome serve the virus, or rather, serve the global process of genetic information realization as a part of the integrated life system.

While not all viruses cause diseases, some are highly pathogenic and cause very serious ailments such as influenza, smallpox, parotitis, poliomyelitis, and HIV infection. Some viruses can infect bacteria and kill them. These are called *bacteriophages*. In a mechanism akin to viruses, the chromosomes of various plant cells can be parasitized by *viroids*—very small bare molecules of annular RNA that consist of just a few hundred nucleotides. These tiny RNA molecules essentially represent a minimal manifestation of a living body which has both a phenotype (a certain quantity of nucleic acids and an annular structure) and a genotype (a certain set of genes).

The origins of such a specific parasitic form of life are probably connected with the early stages of prokaryotic evolution. Certain nucleic acids of nascent cells are likely to have originated plasmid-like 'beings' that adapted and proliferated using cellular resources for their own autonomous existence and reproduction. After certain evolutionary pressures, these beings gradually gained the ability to penetrate other cells and exist in the external environment. Thus, viruses are derivatives of cells rather than independent organisms. Let us examine several considerations to support this claim.

Viral genes have a cellular origin, because they encode proteins that are common for both cells and viruses, but they are able to reproduce and develop only inside cells. For example, bird flu viruses contain neuraminidase, an enzyme which demolishes the glycocalyx of epithelial cells in an animal's respiratory tract. This activity is understandable, because these are the cells in which the cycle of formation for this particular virus occurs. Many viruses are also known to have capsules that are derivatives of the cellular membrane of the host. During their lifetimes within cells, viruses may mutate, evolve, change their properties, and alter their pathogenic sites (e.g., swine flu). Viruses are very selective in choosing which cell types to infect. Their penetration only into certain cells of a particular lineage leads to a successful production of multitudes of new virus particles, which can directly or indirectly result in the illness attributed to the virus. Lastly, although there is a similarity between viral and eukaryotic genomic elements called introns and transposons, the virus requires cellular machinery to realize its functions. It is therefore obvious that despite significant phenotypic differences, both viruses and cells have some common molecular foundations and mechanisms, which testifies to their genetic affinity.

But viruses do not possess one of the main characteristics of living beings—autonomy. Viruses depend totally on cellular structures and processes for their activities. Their molecular compounds correspond in many respects to those of their potential hosts. It can be surmised that cells began to 'produce' viruses at a certain period of evolution. The absence of sexual processes in prokaryotes and ancient eukaryotes could have been 'deliberately' compensated by viral transformation. Subsequently, some population of the viruses went out of control and gained the ability to reproduce independently of the organism they were formed by. This was very likely at a time when only prokaryotes existed, which utilized derivatives of their own genomes as a way of exchanging genetic information. In other words, viruses and their analogues (viruses, viroids, phages, and plasmids) formed as the global mechanism of transfer and exchange of genetic information between the genomes of various cells and within an individual cell as well (transposons, IS elements). Mobile genetic elements unite all discrete genomes into the *integrated information space*, where any NA segment can be transferred to any part of the global genome. Thus, viruses and their analogs are the instruments of the infinite evolutionary process (see Fig. 2.6).

Viruses are very labile. They evolve constantly in concourse with the development of living organisms. At the present time, many mobile genetic elements are probably the competent representatives of discrete genomes of all the cells and organisms that inhabit the Earth. For example, according to different sources, between 8 and 20 % of ordered nucleic acids in the human genome are similar to those found in retroviruses. This leads to the assumption that many viruses primarily exist as *episomes* within cells, while outside they are found in a transitional form as peculiar cysts (see Fig. 2.7).

It is interesting to note that cells on their own are not too strongly opposed to the viruses living inside them, since they do not seem to make any type of protection against them and even have special corresponding receptors. The exact

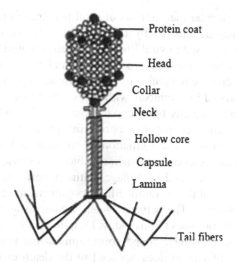

Fig. 2.6 The structure of a phage gives evidence that it is a special instrument for manipulating the molecules of membranes and DNA. The phage attaches to the surface of bacteria by its tail fibers. The hollow core then penetrates through the membrane and forms a channel through which the viral DNA is injected into the cellular cytosol

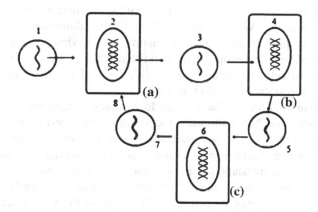

Fig. 2.7 Stages of viral existence—the stages of transformation and transitions of information. 2, 4, and 6: stationary stage (episome) in the genome of one of its hosts. 1, 3, 5, and 7: transitory stage in the external environment. (*1*) Virus in the external environment. (*2*) Virus as part of a human cell genome. (*3*) Mutating virus with 'stolen' segments of host DNA. (*4*) Virus inside the genome of a chicken cell. (*5*) Repeatedly mutated virus that contains segments of chicken DNA. (*6*) Virus inside the genome of a pig cell. (*7*) Repeatedly mutating virus that contains segments of chicken and pig DNA. (*8*) Further infection of human cells. This circulation of viruses among several genomes of different hosts allows it to deceive the protection systems of these organisms. (*A*) Human cell. (*B*) Chicken cell. (*C*) Pig cell

correspondence of molecular mechanisms of viral penetration into a cell suggests that these mechanisms are created by the cell itself. However, a multicellular organism is generally opposed to viral life inside itself, because there is a threat of destruction of certain body parts due to the abnormal functions of the contaminated cells. Long before emergence of multicellular organisms, single-celled creatures probably existed in symbiosis with viruses and their analogues, and this is likely to have been a mutually beneficial coexistence. The cells allowed viruses to reproduce, and in turn they got the opportunity to use the additional genetic material procured by viruses. The formation of colonies and, later, multicellular organisms, led to a contradiction with such symbiosis, because the penetration of some pathogenic viruses resulted in damage to the integrity of multicellular systems. In response to such a threat, multicellular organisms have developed special immune cells for protection. The battle between organisms and viruses has lasted for a billion years, and continues to this today. Despite this, however, many viruses and multicellular organisms have found compromises that imply that the development of a majority of viruses does not lead to the death of the host organism, while in turn the immune system of the host cannot cause significant damage to viruses.

We have shown that viruses are a legitimate part of living nature's global genome system. In our opinion, viruses are the *discrete units of genotypic life—* bodies that thrive at the molecular level. Their life is invisible to the naked eye, and may only be discovered in cases where pathogenic parasitism causes phenotypic changes in a host. We may also conclude that the definition of 'non-cellular forms of life' is quite relative, since the main stage of viral existence occurs in cells. This expression relates only to the definition of the existence of inactive viral bodies outside of cells, because every possible form of life has a cellular basis, including the manifestation of viral activity.

As inhabitants of the molecular world, viruses demonstrate the minimal biological body (one NA molecule), the minimal qualitative and quantitative composition of a living body (NA molecule + several proteins), and the minimal set of processes and mechanisms necessary for living (replication, transcription, reverse transcription, translation, catalysis, and self-organization).

Thus, viruses are probably the ancient derivatives of single-celled organisms that reigned on Earth during the first 1.5 billion years of life's existence. Then and now, their main purpose is the varied transfer of genetic information within the *integrated information space* of the Global Genome (see Fig. 4.8).

## 2.8  Informational Basis *(A more detailed discussion of this issue can be found in Part 4)*

It is known that living organisms are complex open systems that exist on the basis of energy and matter exchanges with their environment. However, an exceptional significance has lately been attached to the role of information on the basis of which organisms are born, survive, improve, and interact with the environment. Here is a list of the main informational flows found in biological systems:

1. *Flow of External Information.* Organisms exist in an 'ocean' of information from the external material world. Absolutely everything around us carries information. It could be in the form of various physical, chemical, and biological phenomena, various types of movement and changes, diverse waves, fields, corpuscular streams, and so on. Living bodies perceive a certain part (a very limited part) of this external information. Organisms are only able to process the information they can sense through special analysers, receptors, and internal systems which can understand and apply the information. Such a specific system of perception and informational analysis in living beings is called the *thesaurus*. The term 'thesaurus' is used in the theory of information for the identification of a package of all the information that a given individual has in its possession. Organisms of different types and species exhibit differences in the way they organize their thesaurus, therefore possessing unique distinctions in perception as well as in the quality and quantity of the information processed. Organisms adapt and live in those informational conditions which correspond to their unique thesaurus in such a way as to ensure their survival.

2. *Flows of Intracellular Information*

   a) Genetic Information. Genetic information is stored in DNA and RNA. During realization, information provides for the synthesis of the necessary proteins. The proteins then participate in their respective cellular functions. The transformation and transmission of information is provided by the processes of replication, transcription, translation, and expression. This type of information is used for reproduction, development, and support of the structural and functional organization of living beings.

   b) Molecular Information. Intracellular structures, as the elements of a complex system, function in an interrelated and synchronous way due to their regulation by biologically active molecules. These molecules carry signals for activation and cessation of metabolic processes, mass transfer, enzyme activation and inhibition, etc. For example, there are several specialized protein factors that regulate protein synthesis, and the end products of biochemical pathways are usually allosteric regulators of key pathway enzymes.

   c) Information in Orderliness. It is obvious that the orderliness of the intracellular content itself constitutes a complex informational system. Inherited by means of cell division, this predetermined orderliness of the cytoplasm defines info-structural conditions for the behaviour of standard biochemical and biophysical processes.

3. *Flows of Intercellular Information.* The body of a mammal consists of hundreds of different types of cells. They form diverse and specialized groups that have specific spatial locations. These cells form tissues, organs, and their functional

elements. A vast amount of cells in a multicellular organism require informational exchanges for the coordination of various metabolic and physiological
processes, division, growth, etc. Several distinct forms of intercellular
communication are presented below:

a) Distant Chemical Signalling. This occurs through liquids of the internal
   environment by means of active molecules secreted by special cells. For
   example, in the case of endocrine regulation, hormones are carried by the
   blood and affect cells that are located in different parts of an organism.

b) Contact Chemical Signalling Between Proximal Cells. This occurs by
   means of special signalling molecules. In this case, cells secrete chemical
   mediators locally, and they are absorbed and used up so fast that they have
   time to affect only the producer cell or nearest-neighbour cells. Signalling
   can occur through direct cellular contact by means of gap junction channels
   formed between neighbouring cells. These channels are the only known
   direct method of intercellular communication, and are variably expressed in
   nearly all animal cells and tissues. Among many vital roles, gap junctions
   are extremely important for proper heart contractions, as they allow all the
   cardiomyocytes to maintain a close functional contact that provides
   synchrony and power. Likewise, during synaptic transmission, neurons
   secrete mediators into special intercellular regions called synapses. These
   chemical substances traverse a very short distance and affect only neighbouring pre-synaptic target cells.

c) Electrical Signalling. This function in the neural system, where signal
   passage within a single cell (neuron) is regulated by electrical mechanisms
   and the transfer of excitation from cell to cell may occur through chemical,
   electrical, or combined means.

In all cases, chemical substances are the carriers of information. They are all
molecules with a definite structure. Even for neural tissues which generate and
transmit electrical signals, it is typical that the final stage of communication with a
target cell is mediated by chemical means as a result of electrical propagation.
After completing their roles, the signalling molecules quickly become altered or
degraded. Not all of the cells in a multicellular organism respond to chemical
signals, but only those that have special receptors. These receptors associate with
the signaling molecule and cause a directed response, leading to the signalling
cascade. The various pathways of informational exchanges via intercellular
communication provide a well-coordinated system for the billions of separate cells
of an organism to act as a single entity.

4. *Flows of Information Between Organisms*

a) Between Individuals of the Same Species. Various concepts of communication also exist between individuals of the same species. They may be
   based on certain behavioural standards, forms, particular postures, facial
   expressions, signs, scents, emissions, sounds, chemical signals, etc. Through
   basic informational exchanges, animals of the same species coexist well in

their territory, survive, and reproduce. More complex methods of recording, storing, and transferring information appear at the level of human beings by the use of language.

b) Between Individuals of Different Species. Many communicational elements can be common to several different species. These may include sounds, movements, or other actions. Many prey animals, for example, possess an inherent ability to recognize their predators by their appearance, scent, or behaviour. Many animals understand the 'language' of marking a habitation territory with excrement or secretions from special glands. The flows of information between individuals of different species direct their co-existence and co-dependence, leading to the establishment of different ecosystems.

5. *Flows of Hereditary (Genealogical) Information.* Besides the intracellular flow of genetic information, life is associated with the use and storage of information within the changing generations of cells and organisms. In this case, genetic information becomes hereditary. Analogous to the way the flow of information is directed from the DNA of one cell to the DNA of daughter cells, it is also directed from one organism to another. This flow is connected with the process of reproduction, and ensures the long-term existence of populations of cells, organisms, and species.

6. *Intragenomic and Intergenomic Information Flows.* Intragenomic transformation and transfer of information occurs during reproduction, and is associated with the infinite number of possible integrations of a father's and mother's genome, the great number of variations of DNA recombination during meiosis, and many variations of chromosome disjunction.

The complex of discrete genomes of living organisms composes the integrated system of the global genome. Though separate genomes belong to representatives of diverse species and vary significantly in their qualitative and quantitative compositions, all of them have the same origin and common concepts of organization and function. Therefore, structural and functional ties still exist, even between distant genomes. Integrating factors for all the discrete genomes include viruses, phages, plasmids, transposons, etc.

We can conclude that there is evidence of continuous informational exchanges at all levels in the organization of biosystems. The first four flows of information we mentioned provide for the life of material bodies (phenotypic life), and the fifth and the sixth flows provide for the continuity of life (genotypic life, or, in other words, life as a natural phenomenon). Living organisms are therefore structured on the basis of information, exist in tight informational surroundings, and live and survive through the ability to generate, perceive, analyse, and use the information around them.

# Chapter 3
# Origin and Development of Life

## 3.1 The History of Life

Life has already existed on Earth for 3.5 billion years—just one billion less than the Solar System itself has existed. The process of the origin and historical development of life on Earth is called phylogenesis. One of the main scientific hypotheses states that life originated from non-living nature due to evolution—a progressive complication of molecules and their systems under geophysical conditions that differed significantly from those prevailing today. By examining fossils, carrying out comparative studies of modern organisms, cells, and molecules, and with the help of scientific modelling, several conclusions can be drawn regarding the basic periods in the origins and development of life. Each of these periods can be characterized by qualitatively new properties of matter:

- Small organic molecules (amino acids, fatty acids, carbohydrates, and nucleotides) formed in the aqueous medium of the oceans in an anoxic, high-temperature environment, under the influence of atmospheric electricity from compositions of core nonorganic substances ($H_2O$, $CO_2$, $CH_4$, $NH_3$, $H_2$) as a result of various chemical and physical processes. At this stage, a new group of compounds appeared called organic substances, which are based on molecules of carbon, and which possessed new properties with respect to their precursors.
- Macromolecules such as proteins, nucleic acids, carbohydrate polymers, and complex lipids formed from the basic organic substances by means of polymerization. These compounds had a set of properties necessary for the construction of complex spatial structures and the ability to store and transmit information.
- Ordered colloidal systems arose from the self-organization of the macromolecules, which united and cooperated to form stable complexes. These systems provided a resistance to the environment and the ability to maintain their internal organization and control the flow of matter and energy within themselves.

G. Zhegunov, *The Dual Nature of Life*, The Frontiers Collection,
DOI: 10.1007/978-3-642-30394-4_3, © Springer-Verlag Berlin Heidelberg 2012

- Certain RNA macromolecules acquired the ability to perform catalysis, regulate protein synthesis, and reproduce themselves by means of complementary matrix duplication. These events led to the emergence of the rybozymes—the first carriers of nucleotide mechanisms for recording information about the structure of molecules. At the same time, ribozymes possessed catalytic properties. On this basis, double-stranded DNA molecules appeared later on, and became the primary nucleotide-based mechanism for information storage and transfer which constituted the foundations of reproduction and development.
- Enzymes appeared as a result of protein evolution. They had the ability to increase the speed of the necessary biochemical reactions by several thousand-fold, arrange flows of energy and matter, and determine the directionality of numerous functions of complex colloidal systems. At this stage, anaerobic metabolism and biochemical energy mediators (i.e., ATP) developed.
- Stable sections with specific nucleotide sequences appeared at certain locations within NA molecules, and these became the genes that established accuracy and continuity in the recording and transmission of genetic information. Further evolution of nucleic acids and enzymes within the colloidal systems led to the appearance of the molecular mechanisms of transcription and translation. From that moment, the primal functions of genetic information were performed by DNA, while RNA became an intermediary in the realization of information: DNA -> RNA -> protein. This allowed the forming organisms to regulate their own molecular composition, structure, metabolism, function, and reproduction by means of the synthesis of necessary proteins. It is reasonable to say that life arose at this stage in the form of long-living bodies—ordered colloidal matrices with an integrated NA processor system. These bodies were capable of self-maintenance and self-reproduction. The origination of protobionts (pre-prokaryotic cells) had occurred.
- Molecular systems capable of self-replication and self-catalyzation became isolated from the environment. Membranes and complex surface apparatuses formed around the protobionts. This allowed the first organisms to acclimatize to their surroundings, become relatively independent from the environment, and better maintain a constant internal composition. Microspheres became stable and autonomous, yet sensitive to the environment. The moment when membranes allowed separation from the environment can be considered as the moment when the first cells appeared.
- The first, very simple prokaryotic cells arose. These are known as archaebacteria. Even at their earliest stages of development, they already had all the general features of living bodies: autonomy, a definite form and structure, centralized genetic material, cytoplasm with enzymes, the ability to maintain complex organization for a long period of time, and the ability to propagate. Some of them live even today, just as they did approximately 3.5 billion years ago (Fig. 3.1).
- Certain archaebacteria and bacteria gained the ability of autotrophy, that is, the ability to synthesize organic molecules from nonorganic ones ($CO_2$, $H_2O$, $NH_3$, etc.). Chemosynthesis was an ancient type of autotrophy where the energy

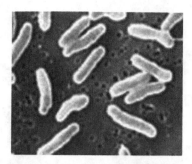

**Fig. 3.1** This is what the first inhabitants of our planet looked like. Not much is needed for successful life. It is enough to have just an elementary autonomic body and a program for its survival and reproduction. Archaebacteria are true virtuosos in modulating their metabolic activities and adapting to extreme conditions: they can switch between anaerobic and aerobic metabolism, withstand high temperatures, up to 100 °C, or on the contrary, freeze down to −196 °C, and survive in highly acidic environments

obtained from oxidation of minerals (iron, manganese, and nitrogen) was used to construct organic substances. Later, prokaryotes developed photosynthesis—a process where sunlight acts as the source of energy. This was a qualitative leap in the course of the development of living nature, because it allowed the use of the unlimited source of solar energy together with the unlimited source of carbon in the form of $CO_2$.

- The first microorganisms capable of oxygenic photosynthesis, called cyano-bacteria, emerged as the result of the evolution of photosynthesizing prokary-otes. They used a readily available substance—water—as a source of electrons and protons, producing photosynthesis by-products such as oxygen. A large-scale process of oxygen accumulation began in the Earth's atmosphere. The ozone layer formed around the planet, and this gave protection to the living organisms from the harmful impact of the severe, ionizing radiation of space, providing an opportunity for the further evolution of life.

- More and more new metabolic pathways appeared, favouring phenotypic diversity and adaptation. It is considered that it is the cyanobacteria that created the conditions in which more complex types of organisms were able to develop. For the first two billion years, the living world consisted exclusively of huge systems of various bacteria. Their impact on the environment resulted in the appearance of ecological, thermal, and atmospheric conditions which were favourable for the emergence and evolution of superior life forms. It is also considered that, in the history of life's development, up to 90 % of all species that have existed have become extinct. The global bacterial system, however, has survived, and continues to regulate the conditions for life on Earth even today.

- Accumulation of oxygen in the atmosphere, water, and ground, and further evolution of prokaryotic cells resulted in organisms that could use oxygen for oxidation and energy production from nutrients through a process called aerobic

**Fig. 3.2** One of the early
multicellular organisms, the
genomes of which have
survived to this day. Picture
of a hydra and its cells. At
this stage of complexity,
genomes gained functions of
differential expression,
conditioning the poly-
phenotypical nature of cells.
During this period, the true
abilities of a genome were
realized in order to build
complex colonies of various
sizes and forms through
regulation of the dosage and
direction of information
distribution

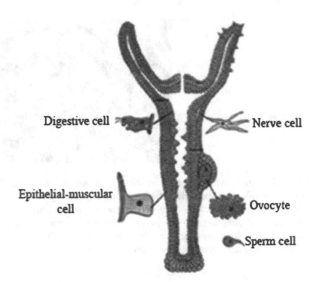

respiration. This highly effective metabolic fate significantly expanded the
energetic facilities of the organisms, and this resulted in the appearance of
various new functions, organizational complications, adaptation, and evolution.

- Continued evolution of genotypes and phenotypes of certain prokaryotic species, coupled with the symbiosis of aerobic and anaerobic organisms, resulted in
the establishment of the first eukaryotic cells (approximately 1.5 billion years
ago). These cells possessed a complex organization and specialized organelles,
and performed a great many functions. The presence of mitochondria and
chloroplasts, for example, provided eukaryotes with tremendous energetic
advantages in adaptation and propagation. They carried a considerable amount
of genetic information enclosed in the cell nucleus. Compartmentalization of
genetic material and supporting enzymes in the nucleus created favourable
conditions for the prolonged storage and optimal functioning of genomes, and
provided accuracy and stability during their reproduction. The alternation of
billions of generations of highly complex organisms has had virtually no effect
on their phenomes or genomes. For example, many insects have existed for
hundreds of millions of years without any significant changes.

- Approximately 600 million years ago, as a result of the continuous evolution of
eukaryotic cells and their colonies, primitive multicellular organisms appeared
(see Fig. 3.2). The association and cooperation of cells resulted in numerous
advantages such as an increase in the efficiency of nutrient processing, the
ability to resist unfavourable conditions, an increase in the survival rate, and the
ability to carry and utilize large amounts of water. These advantages brought
the potential to develop new ecological niches and habitats, e.g., emergence on
land. As a result of progressive development, specialized cells, tissues, and
organs arose on the basis of differential gene expression. This allowed for the
origination of millions of different living bodies that differed significantly one

from another as a result of the needs of the various organisms to adjust to different environmental conditions. After many millennia of selection and evolution, new fundamental and specialized features such as eyes and ears, and nervous, immune, and other systems were created. The evolution of multicellular organisms in various conditions on Earth caused the emergence of millions of species of living beings that have populated all possible ecological niches on our planet.

- Approximately 500 million years ago, evolution of genotypes and phenotypes, compounded with processes of natural selection, determined the appearance of chordates, and later the further appearance of vertebrates, which had significant advantages in structure and function. During their continuous evolution, chordates mastered terrestrial environments (amphibians) and homoeothermy (birds and mammals), and updated all their functions. Mammals came into existence alongside the vertebrates approximately 200 million years ago, and possessed considerable functional superiority in comparison with all the other organisms.
- The complication of the highly developed brain led to the development of complex social structures, sophisticated forms of nurture, and elaboration of individual identification. This basis gave the opportunity to form the specific characters of the most 'intellectual' species that inhabits the Earth—*Homo sapiens*.

The described phylogenic and evolutionary processes continued for several billion years, eventually leading to the emergence of humans. This only occurred in the last few tens of thousands of years.

The above-mentioned general occurrences in the origination and progressive development of living organisms explain the uniformity in structural concepts, molecular composition, and commonality of metabolic fates of all the millions of species of organisms that have ever existed. Changes in geophysical conditions force organisms to continuously re-adapt to new environmental states, which thereafter causes further environmental modifications. Thus, it is apparent that there exists a co-dependent relationship between the environment and the organisms which live in it. By means of evolution, the environment forms the organisms, and the organisms, in turn, form the environment.

## 3.2  What Really Happened?

The phenomenon of life emerged at a certain stage of development of the Earth. It happened as soon as conditions started to correspond to the ability of protein-nucleic bodies to exist stably in the liquid aqueous medium. Probably, over a period of several million years, a network of many diverse phase-solitary colloidal bodies emerged on the basis of the self-organization of dissipative structures. They contained various complexes of NA and proteins, and possessed new properties and merits. The penetration of NA into the protein colloid set the necessary conditions for the emergence of life.

In the beginning, life was a sort of genotypic manifestation without phenotypic features, represented by some amorphous colloidal protobionts that contained NA and enzymes. Later, under certain conditions, various combinations of NAs and proteins started to improve their surroundings by developing phenotypic frameworks, which allowed them to adapt and survive, and which varied in accordance with environmental pressures. These events led to the formation of complex living systems, creating phenotypic life, which was represented by various cells or autonomous living bodies, these being the intermediaries between the genome and the environment. In this way, life emerged as we currently perceive it—through the existence of living bodies. We only notice the replacement of representatives of phenotypic life, the physical intermediaries between the GG and the changing outside space. Meanwhile, genotypic life exists continuously, as it has from the moment of the emergence of protobionts, through the process of molecular evolution, which is hidden from our eyes.

It should be noted that the described stages of evolution of primal living bodies are mostly connected with the attempted attainment of perfection in the organization of structures and processes. Once established, the pattern of organization of cells was cloned and propagated for many years along with the DNA and cytoplasm during the processes of reproduction. In other words, a fundamentally new system was created billions of years ago: a cytoplasmic colloidal matrix (operational unit) with a built-in genome (memory and processor). Not even the following billions of years imposed any significant changes or complements to this order. Evolution has created a tremendous amount of variation in living bodies just through the modifications of qualitative and quantitative molecular compositions of the same system (e.g., bacteria, plants, animals, etc.). Every kingdom of living organisms contains various differences in composition and in the order of intracellular systems of molecules.

Although this process allowed the existence of millions of variations of cells and organisms as well as their respective mechanisms and functions, all of them originated from the same organizational patterns and reproductive mechanisms that were created by nature many years ago.

It would be impossible to understand the essence of life without realizing the process of its emergence and development over billions of years, as described schematically above. With all the continuous changes on Earth, what has actually remained unchanged since life's conception? Probably, it is only the molecules of RNA, DNA, and proteins, and the universal mechanisms of their interaction in the aqueous-colloidal medium. Molecular mechanisms, such as replication, transcription, and translation, still constitute the basis of life's process within every species. That is, throughout the entire existence of life, only phenotypes have been changing (combinations of interacting molecules and cells), while the molecules of DNA, for example, have experienced only atomic and combinatorial modifications.

Life emerged under certain conditions (temperature, pressure, presence of the atmosphere, absence of oxygen, etc.), through the global spontaneous process of the self-organization of moving matter as a phenomenon of the material world.

The essence of this phenomenon was the appearance of the system of the long-lived global genome, together with its attendant processes. The qualitative transition of matter in the process of self-organization was connected with the appearance of living bodies, which were capable of interacting with the environment, using its substances and energy to maintain genomic homeostasis.

Therefore, life is one of the directions of material development that emerged at a certain stage of natural evolution during certain physicochemical prebiotic conditions. It centred around the genome, which is the substrate of life that is regularly ordering material space around itself.

Life evolves and changes almost simultaneously and in correspondence with the modification of the physicochemical factors of Earth as its integral part. This phenomenon will exist as long as the physicochemical conditions for the 'nucleic-protein-aqueous' foundations of life prevail.

## 3.3  Life Derives From Life

Just one and a half centuries ago, it was considered that the appearance of life is possible from non-living bodies and substances. Despite numerous attempts by different scientists to artificially create even the simplest organisms, no one has ever succeeded. It was therefore determined that only living organisms can reproduce themselves in short periods of time relative to the amount of time it has taken for life to establish spontaneously on Earth. This phenomenon of life deriving from life is called biogenesis. Biogenesis is realized through the process of reproduction, where representatives of the same species produce similar individuals that are almost identical to the parental organisms in morphological, physiological, behavioural, and other characteristics (homogenesis). *Mechanisms of reproduction are described in* Sect. 9.1.

The basis for the multiple accurate reproductions of cells lies in the conserved properties of hereditary materials. Every species has an exclusive set of DNA molecules which contain a specific composition and special combinations of genes. The DNA molecules are duplicated and checked for integrity, and then divided between the parent and daughter cells. In this way, the newly synthesized cells appear as exact copies (clones) of their parents. This mechanism goes on for many generations and gives species (or rather, genomes of species) the opportunity to exist for many millions of years, in contrast to the much shorter lifespans of individual representatives.

# Chapter 4
# Discreteness, Order, Organization, and Integrity

## 4.1 Principles of Organization

Organisms consist of many components that are strictly ordered in time and space. The organized interactions of these components establish conditions for the integrity of life. In our view, discreteness and organization should be singled out in the first place as the fundamental qualities of living systems.

*Discreteness*—is the intermittence in the organization of material bodies. The discreteness of an organism's structure means that the organism, as a system, consists of many separate, complex, interrelated, and interacting parts. Discreteness is a very important property of any system, since it provides internal movement and interaction of systemic parts, and therefore provides for the manifestation of its various merits, features, and functions.

*Organization*—determinative principles of the arrangement and functions of systems and the natural laws of ordered relations. The organization of living bodies is rooted in principles of rational structure formation and purposeful functioning of cells and multicellular organisms. It also depends on ordered purposeful relations, interconnections, interactions, and concepts of management of the components of biological systems:

1. We suggest that every organism possesses *unity in the interactions of the genome and phenome* (as proposed by B. Mednikov in 1982), which is one of the basic organizational concepts of living bodies. In other words, each organism appears to be a genetic program for reproduction, development, organization, function, and management, as well as the end product which is manifested as a certain body. The expediency of such unity is guaranteed self-preservation and reproduction.
2. The necessity of the *systemic concept of organization* is related to the appearance of absolutely new qualities and properties of systems in comparison with their constitutive elements, and also the appearance of abilities to form various systems of interacting elements in order to implement numerous functions.

G. Zhegunov, *The Dual Nature of Life*, The Frontiers Collection,
DOI: 10.1007/978-3-642-30394-4_4, © Springer-Verlag Berlin Heidelberg 2012

Hierarchy, total interconnection, and dynamics are the foundation for the existence of living systems. All living bodies are open heterogeneous systems in non-equilibrium, with properties dependent on the qualitative and quantitative composition of individual elements. We believe that the concept of the *stable unbalance of systems* is one of the main concepts of a living system's organization.

3. The *autonomy principle*, which is manifested in the isolation of cellular contents from the environment, is a significant organizational basis for the existence of living bodies. Therefore, organization-wise, living bodies are loop systems. The behaviour of purposeful processes is only possible in the limited domain of organized molecular systems of a particular qualitative and quantitative composition.

4. The life of a body is a *process*. Life is permeated with constant interactions and continuous shifts between states of elements within biosystems. Two groups of rational and purposeful acts occur simultaneously:

   a  A complex of processes aimed at self-preservation.
   b  A complex of processes for reproduction and development.The realization of *processuality of life* promotes an understanding and recognition of evolution.

5. One of the most significant organizational principles of living bodies is their constant *interaction with the environment* by means of matter, energy, and information exchanges. Such interactions emphasize the inseparable connection between living bodies and their environment. The expediency of such interactions is related to the constant consumption of matter and free energy for self-preservation, propagation, and development.

6. The *principle of self-organization* means that the life of an organism is conditioned by the purposeful transformation of substances and energy on the basis of *selective catalysis*, forming a complex self-maintained system.

7. The *principle of informational organization* consists of the fact that genetic information appears to be the directing force of all processes and the means of organization and management of biosystems.

*Organism*—This is a complex isolated system that consists of numerous subsystems, complexes, and molecules. All elements of the system are interdependent and interconnected. This type of system is represented as a complexly organized unity in the form of an integrated organism. Therefore, order and organization of discrete elements of a system leads to a qualitatively new result— the emergence of the integrity of living bodies. As a unit, such bodies respond to various irritants, move, reproduce, possess a standard set of biochemical reactions, and perform various specific functions. Such integrity is determined by the strict hierarchy of structure, cooperation, and coordination of all components by means of neural and endocrine regulation, intercellular communication, and other vital processes.

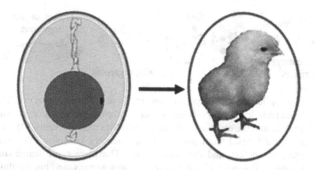

**Fig. 4.1** The 'miraculous' transformation of matter in a closed space in 400 h. In this example, an extremely complicated organism is being constructed from a chaotic assortment of organic molecules and water. 40 g of a non-living, non-organized colloidal solution of organic substances in the enclosed space is transformed into a practically equivalent mass of organized living cells, organs, and tissues. Such an organism is now capable of independent motion and nourishment, and possesses specific functions and behaviour. The organizational *center* of such a fantastic transformation is the tiny embryonic genome

*Order and Orderliness*—A naturally determined arrangement of material bodies in space. This feature is extremely important for living systems because, in a particular space and at a particular time, the necessary elements of a system assemble in such a way that their appropriate disposition and ordered interactions condition the appearance of qualitatively new properties of the given system, despite the second law of thermodynamics. Such a system has clearly delineated boundaries that define its shape and volume.

In nature, the majority of substances are spread chaotically in approximately equal ratios. However, living organisms contain certain substances in concentrations that are thousands of times higher in comparison with the environment. In particular, despite being microscopic, a cell contains a high concentration of various proteins, nucleic acids, specific lipids, and carbohydrates, which are almost absent in the environment. Moreover, the molecules of these components are highly ordered and organized in a particular way within the cell, forming complicated specialized structures. Due to such a strict selectivity and order of molecules, cells form functional structural units called organelles. These organelles determine the functions of different cells, which then arrange themselves in various combinations to form specialized tissues and organs, whose further arrangement establishes the integrated organism.

The organization and order of a multicellular body is created from the disarray of the environment. This occurs gradually, during developmental processes, on the basis of genetic programming, since it is the DNA molecules that constitute the organizing factors in living organisms (see Fig. 4.1). The principles of organization and order are not created every time, but are constantly being copied. For example, one ordered cell originates two, then four, then eight, and so on, all starting with the cloning of DNA, then RNA, and finally proteins. Mechanisms and

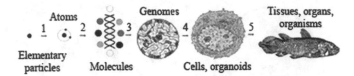

**Fig. 4.2** Stages of complication of biosystems during the development process. The integration and interaction of elements leads to the progressively complex growth of a system, qualitative steps in its development, and the emergence of new properties and functions. (*1*) and (*2*) Specific atoms and molecules are formed from nonspecific elementary particles. (*3*) These molecules make up membranes, organelles, and the genome. (*4*) The phenotypic framework is formed around the genome—the cell is born. (*5*) Body parts, tissues, organs, and multicellular organisms are eventually established

processes are copied along with the proteins of the cytoplasm and karyoplasm, and the functions of proteins maintain the cloned order.


## 4.2  Levels of Life

Discreteness is the foundation of the series of structural and functional levels of organization of living organisms and biosystems (see Fig. 4.2). Each level has greater complexity and possesses new properties and functions. It will henceforth be convenient to identify the following organizational levels of life:

1. *Elementary Particles.* These are protons, neutrons, electrons, etc. Various combinations of such particles form atoms of different elements.
2. *Atoms.* Depending on the combination and quantity of elementary particles, they form all the diversity of elements on our planet. Some of them, like carbon, oxygen, hydrogen, nitrogen, and phosphorus, compose the majority of the organic world. Atoms form molecules by associating with each other in different combinations and amounts.
3. *Molecules.* All living organisms consist of up to 70 % of water molecules. The rest of the composition belongs mainly to the four major groups of organic molecules: proteins (up to 20 %), carbohydrates, nucleic acids, and fats. Energy is transformed and stored as high-energy bonds of multipurpose molecules of ATP. Hereditary information is stored and implemented by molecules of DNA and RNA.
4. *Organelles.* These are formed as a result of the organization of cellular macromolecules. This results in the formation of specialized cellular structures such as biomembranes, mitochondria, lysosomes, ribosomes, etc. Various organelles have a specific macromolecular composition that conditions the peculiarity of their structure and function. The complex of organelles acting in coordination forms the cell.

5. *Cellular Organism.* Cells are the units of life and are typical for the organization of all organisms. Metabolism, homeostasis, biosynthesis of proteins,
   realization of hereditary information, and reproduction are possible only at the
   cellular level. Through the process of evolution, the first colonies of similarly
   structured cells appeared, grouped together, and gained specialized functions,
   later becoming the tissues of multicellular organisms.

6. *Tissues.* These are complexes of cells with a similar type of organization and a
   particular set of functions. Hundreds of different types of cells compose bodies
   of different multicellular organisms. In animals, numerous cells form 4 primary
   types of tissues: neural, connective, epithelial, and muscular.

7. *Organs.* These are highly modified body parts, situated at well-defined locations and possessing specialized functions. They are formed during developmental processes from the cells of various tissues. Higher order animals have
   many organs, which are diverse in size and structure and perform many
   functions.

8. *Systems of Organs.* These are groups of different organs that function in
   cooperation in order to perform common tasks for the whole organism. For
   example, humans have the following systems of organs: digestive, respiratory,
   cardiovascular, neural, excretory, reproductive, endocrine, locomotive, and
   integumentary systems. Each separate organ performs its specific role, but all
   together they work as a 'team'. All systems of organs function interdependently, regulated by the neural and endocrine systems.

9. *Multicellular Organism.* Interdependently functioning cells, tissues, organs,
   and systems of organs form a multicellular living body. This is an organism
   that is a unit and a carrier of a higher level of life. All previously listed levels
   (molecular, cellular, organ, etc.) work in cooperation in order for the whole
   organism to survive. Therefore, a multicellular organism is a super-system that
   consists of numerous molecules, organelles, cells, tissues, organs, and body
   parts.

10. *Population/Species.* This level describes a set of organisms that are similar in
    morphology, physiology, and peculiarities of metabolism, and are capable of
    mating and reproducing their own kind. This is the definition of a classic
    biological species. It is a very successful and widespread form of existence of
    organisms in communities. However, it is not unique. Millions of bacteria and
    protista, incapable of sexual reproduction, form 'clonal species', individuals
    of which are incapable of mating and which reproduce their own kind only in
    agamous ways. Certain groups of organisms have much more sophisticated
    ways of reproducing similar individuals, for example, through so-called
    'species complexes' that are represented by the association of different species
    with individuals capable of mating under certain conditions. This phenomenon
    is found in the world of protista, fungi, and even vertebrates (fish, frogs). The
    Earth is inhabited by several million different species of animals, plants, fungi,
    protozoa, and prokaryotes. Every species has groups of organisms which
    possess certain morphological and physiological peculiarities and which
    assemble into populations.

11. *Ecosystems*. Historically formed, stable communities or large-scale systems of populations that consist of different species, interrelated with each other and with the non-living environment by means of matter, energy, and information exchange.

12. *Biosphere*. The complex of ecosystems assembles to create the biosphere, which is the integration of all life on Earth, incorporating all of the organisms that exist in a tight association with non-living nature in the atmosphere, hydrosphere, and lithosphere. The biosphere consists of billions of different species of animals, plants, fungi, protozoa, bacteria, and viruses. It is the highest level of organization of living nature, and is the interrelated unity of all living organisms.

From the above-listed biological systems, it is convenient to consider only cells and multicellular organisms as being truly alive, because they differ fundamentally from other biosystems by their autonomy and their ability to reproduce without any practical changes in appearance or function. Although these traits are vital, the most important point to consider is that only they are the real owners of genomes. In other words, not all biological systems are alive.

Here we present a summary of the key components in the organization of biosystems:

1. DNA molecules are the units of genetic information and constitute the foundation of molecular organization. Proteins are synthesized from the information stored in the DNA, which contributes to the sequential construction and maintenance of molecular order. This particular flow of genetic information leads to the creation of different cells with all their internal components and organelles.

2. Cells are the units of life and the organizers of all the variable aspects of organisms (single-celled as well as multicellular). Many phenotypes (billions of variations) of cells of animals, plants, fungi, and unicellular creatures are constructed on the basis of different genotypes. Such a great diversity of cells delivers an infinite number of variations of multicellular organisms by means of cellular cooperation in various combinations and quantities and also on the basis of their interactions.

3. Multicellular organisms are the units of organization at the population level in biosystems and, just like individual cells, are the carriers of life.

4. Species are the most basic taxonomic units used to classify similar organisms. On the basis of their structural and functional peculiarities, species are united into greater taxonomic classifications such as genera, families, orders, classes, phyla, and kingdoms of living organisms.

All of the above-listed units in the organization of life are, ultimately, derivatives of DNA. Moreover, each level differs by qualitatively new properties which appear to be the basis for the emergence of the next higher level of organization. The first derivative of DNA is RNA, the second is proteins, the third is cells, the fourth is organisms, and the fifth is species.

The principles of discreteness, organization, order, and integrity refer not only to the structure, but also to the functions of organisms. In particular, the metabolism of any cell is discrete, because it consists of thousands of individual distinct biochemical reactions. At the same time, however, it is also integrated, since it serves a single purpose—the maintenance of homeostasis. All metabolic reactions are highly ordered in time and space. Each reaction goes on in a well-defined part of a cell, implemented in a strict order and at a strictly defined time. The organization of discrete processes is provided by selective catalysis of only those reactions that are necessary for the cell at that time. This type of catalysis is based on the sole presence of the necessary enzymes, which are synthesized by the cell on the basis of genetic programming.

Energy in living systems is also transformed and used on the basis of these same principles. Its provisions are selectively directed to fulfil the needs of only the necessary reactions and processes. These guiding concepts are therefore typical for the specific processes of molecular interpretation and utilization of biological information.

The diverse organizational levels of all aspects of life are created on the basis of discreteness by the hierarchy principle: lower levels are the direct constituents of higher ones, and therefore condition new properties in them. This concept can be extended to various examples of discreteness, such as a scale of sizes and masses, atomic structures, energy levels in quantum mechanics, etc. In other words, discreteness is one of the major characteristics of nature.

From the dualistic perspective, not only phenotypic life (which was described earlier in this chapter), but also genotypic life appears to be discrete, integrated, and organized. The system of the global genome consists of an infinite multitude of discrete genomes from diverse unicellular and multicellular organisms. These genomes are rigorously organized in the composition of autonomic cells, organisms, and species. At the same time, however, they are also interconnected by genetic universality and extension into the integrated system of the circulation of genetic information.

# Chapter 5
# Living Systems

## 5.1 Open, Nonequilibrium Systems

A *system* is a unified singularity which is composed of many interrelated elements that are substantively or conventionally selected. In this chapter we shall discuss biosystems, focusing specifically on the living bodies of cells and multicellular organisms. As previously stated, it should be clear that not all biosystems are living. For example, the population of organisms is a biosystem, but is not itself considered living. The nucleus of a cell is also a biosystem, but once again is not a living body.

The cell is a very complicated system which consists of complexes formed by the interactions of numerous diverse and ordered molecules. These molecular complexes form organelles, which are structurally and functionally associated with each other and act as the components of a cellular system. The organelles of the intracellular space such as the nucleus, mitochondria, lysosomes, etc., are divided by biomembranes into distinct compartments where only the specific reactions particular to each compartment occur regarding the exchange of matter, energy, and information.

As mentioned previously, several levels of organization are also typical for multicellular organisms. From the top down, they consist of body parts, organs, tissues, cells, intercellular structures, fluid media, molecules, and so on. The lower levels of organization condition the structures and functions of the higher ones, which in turn determine control of the lower levels. The vital activity of cells can be performed only under conditions of coordinated interactions between cells themselves and between them and the environment, which facilitates a constant exchange of matter, energy, and information. The processes of cellular interactions allow metabolic and physiological events to occur within these cells and multicellular organisms in a coordinated and ordered manner.

Organisms and cells require constant exchanges of various things with the environment. They need to replace worn out and destroyed parts by new ones and they need to sense external conditions and perform certain functions based on the

G. Zhegunov, *The Dual Nature of Life*, The Frontiers Collection,
DOI: 10.1007/978-3-642-30394-4_5, © Springer-Verlag Berlin Heidelberg 2012

input. Without these types of communication, the high degree of cellular order would eventually be disrupted with time, leading to the collapse of the organism. Thermodynamic systems of this kind which constantly obtain an influx of matter and energy from the outside are called open systems. From a thermodynamic point of view, an organism is a highly organized, open, non-equilibrated system, which constantly interacts with the environment. Such systems transform the chemical energy of nourishment into the energy needed for biochemical and physiological processes. Any unused energy and unorganized matter is either excreted to the environment as various waste products, or is stored for future use (i.e., glycogen).

Systems possess a number of physical characteristics called parameters. For example, living organisms can be described by their temperature, size, volume, qualitative and quantitative composition of proteins, and so on. A complex of parameters, typical for the given system, defines its thermodynamic state. A change in one or several of them is called a thermodynamic process. If the parameters can change reversibly, either spontaneously or under the impact of external forces, such systems are said to be in non-equilibrium, which is a term that describes all biological systems. Thus, they can be modified spontaneously or under the influence of various factors, temporarily changing the process flow, and thereby affecting the condition of the whole system.

Living bodies are peculiarly 'exciting' systems, because their structural organization possesses great internal energy. Such energy is constantly being dissipated during metabolic processes and through the erratic thermal destruction of the system due to the chaotic thermal motion of molecules. This means that biosystems are inherently dissipative in nature. They undergo constant energy transitions during the conversion of ordered events into chaotic ones. In other words, processes in living systems that attempt to order the surrounding space are continuously associated with an inherent increase in entropy (disorder). That is why, in order to survive, biological systems need to maintain their organization artificially by means of instant influxes of free energy and necessary substances. Minor changes in the energy or matter supply may lead to irreversible damage to a living system. Thus, within the varying temperature ranges of life, macromolecules and their complexes are quite labile and unstable. The organization of these living systems can only be maintained through anabolic processes that constantly eliminate chaotically appearing malfunctions connected with thermal and other forms of destruction. Therefore, on the basis of their enclosed genetic programs, biosystems are constantly working against their own destruction.

Even residing in a state of non-equilibrium, organisms are still quite stable systems, that is, their parameters remain unchanged for certain periods of time. As previously noted, this is provided by the constant exchange of matter with the environment and the usage of free energy. In a mature organism, the consumption of matter and energy corresponds to their utilization and excretion. For example, a living system needs to be constantly supplied by components such as water, salts, oxygen, and various organic substances. These substances are vital to the organism and are in constant demand, because it cannot synthesize its own, and therefore would not be able to exist without them. Concurrently, harmful agents such as

$CO_2$, toxic metabolites, and heat should be withdrawn from the organism. Therefore, living bodies should ideally consume ordered matter, while excreting less organized and potentially hazardous 'waste'.

Biosystems are self-organizing due to the purposeful utilization of free energy and matter, which leads to the formation of ordered and functioning structures. The processes involved in the self-organization of biological systems are controlled by genetic programs. This is especially conspicuous during embryogenesis, when genetically determined processes of morphogenesis and differentiation create a highly ordered and complex organism from a single cell within a short period of time.

Living systems are also hierarchic because they have several levels of organization, with each later level being interrelated with and dependent upon the former. Through definite processes, the molecular-genetic level conditions the organization and functionality of cells, and the interactions between these cells further condition the organization and functions of tissues and organs. This iterative hierarchy leads to the formation of an organism in which everything is interconnected in one space and controlled by special systems and mechanisms. Furthermore, biological systems are also cybernetic, because their functions and regulation are based on principles of generation and utilization of information. Cybernetic systems are organized and ordered complexes of interrelated and interacting elements which are capable of generating, sensing, memorizing, processing, and exchanging information. All living bodies sense environmental information, process it, and then respond in a corresponding manner.

Organisms are highly dynamic systems that exist on the basis of the constant motion and work performed by their constitutive elements. In this respect, they are also evolving systems, whose development is stochastic and nonlinear. At certain times, however, the dynamic processes within the organism gain a specific direction, which is defined by genetic programming and selectively catalyzed events.

Living systems possess the mechanisms of self-regulation and regeneration. These mechanisms maintain their integrity and the constancy of their internal environment (homeostasis), which provides for the stability of metabolic and physiological processes and a relative independence from the external environment. Organisms adapt rapidly to environmental modifications by means of these homeostatic mechanisms, which are also controlled by the genome.

Organisms are open systems, but if analysed together with the environment, they form a closed system, where, in general, entropy is increased according to the laws of classical thermodynamics. The law of conservation of energy and matter, in this case, conditions the constant cycling of matter and energy between living and non-living nature. A living being, after the end of its vital activities, returns itself and all of its components back to the non-living realm, so that the general amount of matter and energy remains unchanged. Although individual living bodies may cease to exist, new ordered biological systems take their place, centered around the reproductive and developmental processes associated with the global nucleus of genetic information.

Finally, the systemic concepts presented here are also typical for many other biological consolidations. For example, it is important to note that numerous discrete genomes, unified by their NA origins, form the system of the *global genome*. Likewise, numerous phenomes, unified by the origin of proteins and the commonality of their manifestations, form the system of the *global phenome*. Possessing an interdependent relationship and conditioning each other, these two systems form the *integrated system of life* (ISL).

# Chapter 6
# Life States

## 6.1 Latent Life and Anabiosis

In compliance with the evolution of the Solar System, diverse geophysical conditions occur in different areas of our planet. Primarily, these areas differ in their temperatures and abundances of water.

Life has emerged through the process of continuous evolution of organic molecules and their integration and interrelations in a liquid medium. Because the majority of living organisms consist of about 70 % water, the thermal conditions of the environment must maintain water in a liquid state in order for the molecules to undergo natural motions and interactions. Therefore, the temperature range for the existence of life is generally from 0 to 90 °C. At temperatures lower than 0, water is a solid, and at temperatures higher than 90 °C, irreversible damage is inflicted on the molecules of life: DNA and proteins.

The rate of the motion and interaction of molecules diminishes with a decrease in temperature. During such an event, a corresponding decline in metabolic activity occurs within cells, which results in the inhibition or complete cessation of living processes and functions. The same changes in cells can be observed under conditions of dehydration, since in the absence of water, molecular motion becomes impossible and biochemical reactions can no longer proceed. Thus, in response to alterations in temperature and water, living processes may continue at different rates, become inhibited, or even reversibly come to a halt.

As a result of billions of years of evolution in certain environmental conditions, life on Earth can exist in different states, depending on the degree of manifestation:

1. *Biosis.* A dynamically vital state of cellular activity that manifests all the properties and features of life: nourishment, respiration, excretion, motion, reproduction, etc. Such a state is typical for the majority of organisms that live in normal ecological conditions in temperatures ranging from 5 to 40 °C. Genetic programs, recorded in DNA molecules, are actively realized in this state, maintaining all the necessary processes of life.

G. Zhegunov, *The Dual Nature of Life*, The Frontiers Collection,
DOI: 10.1007/978-3-642-30394-4_6, © Springer-Verlag Berlin Heidelberg 2012

2. *Hypobiosis*. A state of suppressed vital activity, such as that found in cells whose environmental conditions are in the lower bound of the vital temperature range, approaching 0 °C. It is related to the disruption of the liquid state of the internal environment, which would normally ensure the occurrence of metabolic reactions and flows of metabolites. When they approach this lower bound, organisms may switch to the hypobiotic state, which is characterized by an abrupt inhibition of motion, nourishment, respiration, excretion, and other signs of life. For example, many amphibians and reptiles enter a comatose state in cold conditions, and ground squirrels, marmots, and some other mammals go into hibernation, which is a prolonged sleep-like state in the winter. While in this state, body temperatures may drop to 0 °C for quite a long period of time. At this point, such organisms turn on alternative genetic programs, which lead to the turning on of new genes or modification in the activities of normally-functioning ones. These new programs bring the metabolic activities of the organism to another level, and maintain the processes of adaptation and existence in these new and extreme conditions. If normal temperatures are re-established, the standard genetic programs of life are restored, and the organism returns to a biotic state.

3. *Cryptobiosis*. A state of hidden or unobservable life. This is a condition of physiological rest, based on the adaptation that encourages the ultimate survival of the organism in the presence of unfavourable environmental factors. During this state, virtually no signs of life are observed, although extremely low levels of metabolic processes are occurring. Life forms such as the spores and cysts of microorganisms, algae, fungi, seeds and buds of plants, gametes of animals, diapauses of arthropods, etc., are able to undergo cryptobiosis. In this condition, the genetic material is conserved, but there is virtually no realization of the information stored in the DNA molecules, and unchanging forms and levels of DNA and protein may remain for a very long period of time. The resting state is usually preceded by certain preparative events such as hoarding of nutrients, dehydration of cells and tissues, and a decrease in metabolic activity, ensured by special genetic programs. Through this phenomenon, life on Earth persists, develops, and even spreads despite unfavourable environmental features.

4. *Anhydrobiosis*. The ability to withstand extensive and prolonged periods of dehydration together with the ability to fully recover all vital activities upon restoration of water. This phenomenon is typical of certain microorganisms, plants, and invertebrates. As opposed to cryptobiosis, anhydrobiosis is initiated during rapid dehydration without a period of prior preparation. The water content in anhydrobiont bodies may drop to 1–3 %, which is not enough for the functioning of most biochemical processes. Genetic programs cannot be realized under such conditions. The living processes of these organisms can cease for long periods of time, but they resume rapidly after re-hydration. Numerous cyanobacteria (*Stratonostoc*), algae (*Protococcum*), and fungi (*Auricularia*) possess the ability to survive in such a state. Some representatives of bacteria, which are approximately 260 million years old and which have been found in the salt sediments of ancient seas and lakes, possess the ability to revive in favourable aqueous conditions. It therefore appears that as long as the structures

of DNA and various essential proteins remain undamaged, these organisms have the ability to restore their vital processes.

5. *Anabiosis*. A temporary, reversible arrest of life by such environmental factors as deep freezing, deep dehydration, or a combination of these and other events. This phenomenon is typical for certain microorganisms, plants, and simple animals (e.g., rotifers, barnacles, etc.). Some representatives of amphibian and reptilian species are capable of sustaining freezing winter conditions for lengthy periods, up to several months. In essence, they 'die' and then return to life again. Bacteria, which have been found in the ice of Antarctica at depths of several thousand meters, have been successfully revived in laboratories after remaining in anabiosis for tens of thousands of years. Seeds and spores of plants, various microorganisms and bacteria, spermatozoa, embryos of various animals, erythrocytes, bone marrow cells, and many other organisms can be stored in liquid nitrogen with cryoprotectants at temperatures close to $-196\ °C$. In a completely frozen state, these beings possess no vital activities and may remain in such conditions for many years. Their DNA and proteins remain in a cryopreserved state, and any realization of genetic information or metabolism is completely absent. Nevertheless, they may completely re-establish their vitality after thawing and recovering under favourable conditions. In some microorganisms, anabiosis through freezing is also linked with dehydration, which is reversed upon thawing. The basis for recovery from this lifeless state of existence is also grounded in the ability to retain the proper structure and functions of proteins and nucleic acids even at liquid nitrogen temperatures and/or conditions of full dehydration.

Therefore, vital processes may occur at different rates, significantly retard, and reversibly halt. However, in all cases, reversible arrest of life is only possible under conditions wherein the structural foundations of organisms are retained, or, at least, certain undamaged structures are retained, especially the DNA and proteins.

Because there are no visible manifestations of life, can we consider frozen spermatozoa or rotifers as being dead? The answer is no, because they retain the ability to return to normal life after thawing. In other words, the freezing of such living bodies may be considered as a temporary lull in life's processes, or, rather, as a temporary cessation of life's vital manifestations with the retention of structure, organization, and order. This means that phenotypic life is just a consequence of the potential properties of highly-organized matter, as determined by the genome. Such properties might not be manifested if conditions are not optimal. Thus, in a frozen state, living bodies do not possess their normal biological behaviour, but still retain their high levels of order (low entropy) and the genetic potential for revival.

The phenomenon of the temporary absence of active life gives evidence for the possibility of the existence of hidden life in the resulting quiescent DNA and proteins. These molecules—the main organizers of life—may immediately manifest themselves at the return of favourable conditions and display their native processes and characteristics. This leads us to surmise that specific conditions are necessary for the implementation of biological processes and the phenotypic manifestation of life! Meanwhile, such conditions are not needed for the 'silent' existence of life's substrates. In order to exist, therefore, it is not necessary to live permanently. The ability to exist in discrete manifestations is also one of the main properties of life. We may thus say that life *exists* as genomes while it *manifests* itself as phenomes.

When we say 'life exists', we assume, in the first place, that *a genome is present*. The statement 'life does not exist' is the conclusion that this feature is absent. If we say that 'life manifests itself', we mean that there is a presence of *living bodies* that perform locomotion, respiration, and nourishment, involve intracellular chemical and physical processes, and maintain integrity and order within and around themselves. Conversely, if we say that 'life does not manifest itself', this means that the given living bodies do not move, do not breathe, do not feed, and do not perform metabolic processes. This description may be connected not only with the appearance of death, but also with a temporary cessation of living processes. In other words, even if all signs of life appear to be absent, this does not necessarily mean that it is absolutely lacking. The life of an organism, therefore, is the specific manifestation of its genomic potential, which is defined by certain conditions and interactions with the physical world around it.

## Outlines of Duality

In summary of the first part, which describes the most significant characteristics of life, let us note certain outlines of duality.

*Living Bodies vs. the Phenomenon of Life & Genome vs. Phenome*

1. Life is represented by specific living bodies, which are the carriers of life as a global phenomenon. Thus, on the one hand it is a property of *particular physical bodies*, and on the other hand it is *a property of nature itself*.
2. The phenomenon of life is *integral*, though it is represented by the multitude of *discrete units* of living bodies which are physical and phenotypic representations of their genomes.
3. Discrete genomes are united by the common nature of NA into the integrated system of the *global genome*. Discrete phenomes are united by the common nature of proteins into the integrated system of the *global phenome*. Being interconnected and conditioning each other, the two together form the *integrated system of life* (ISL).
4. The supreme expediency of organization within the ISL is due to the correspondence (complementarity) of a system of genomes with the material environment by means of a system of qualitatively diverse phenomes.
5. Life as a phenomenon appeared on Earth as a system of genomes (GG) surrounded by discrete phenotypic frameworks in the form of prokaryotes.

Separate living bodies of protobionts could have emerged earlier, but it was not a global phenomenon until they started to interact, reproduce, and form stable systems, which possessed fundamentally new properties.

6. Evolution is a global phenomenon which is an inherent property of the system of genomes and is realized through the individual interactions of phenomes with the material environment.

7. Discreteness, order, organization, and integrity are typical for both living bodies and life as a global presence in general, although living bodies descend from other living bodies by means of reproduction (sexual or asexual), while the phenomenon of globally interconnected life is permanent.

8. Systems of GG, GP, and living bodies are open biosystems in a state of nonequilibrium, although only the bodies are considered to be truly alive.

9. Time has a completely different meaning for living bodies and for the phenomenon of life. Organisms live only for a certain period, while the phenomenon of life is continuous and infinite.

10. The appearance of a specific protein in its functional conformation is highly improbable, and so is the emergence of life as the phenomenon we observe today. At present, the definition of probability can correspond only to the structures and processes of living bodies, but does not relate to the overall phenomenon of life as a form of material existence.

11. Anabiosis is an excellent demonstrator of the duality of life. On the one hand, there exist living bodies at certain temperatures, and on the other hand, there is a complete arrest of vital activity in a frozen state. After thawing, the bodies become alive again. *The main question is this: where does life reside between the periods of activity of living bodies?*

# Recommended Literature

1. Bernal, J.D.: The Origin of Life. World, Cleveland (1967)
2. Oparin, A.I.: The Emergence of Life on Earth. Moscow (1957)
3. Bendall, D.S.: Evolution from Molecules to Men. Cambridge University Press, Cambridge (1983)
4. Darwin, C.: On the Origin of Species. Murray, London (1859). Reprinted, Penguin, New York (1984)
5. Miller, S.L.: Which organic compounds could have occurred on the prebiotic earth? Cold Spring Harb. Symp. Quant. Biol. **52**, 17–27 (1987)
6. Vernadsky, V.I.: Biosphere (Selected Works on the Biogeochemistry). Moscow (1967)
7. Cech, T.R.: RNA as an Enzyme. Sci. Am. **255**(5), 64–75 (1986)
8. De Duve, C.: Blueprint for a Cell: The Nature and Origin of life. Neil Patterson Publishers, North Corolina (1991)
9. Mednikov, B.M.: The Axioms of Biology. Moscow (1982)
10. Alberts, B., Bray, D. et al.: Molecular Biology of the Cell. Garland Science, New York (1994)
11. Frank-Kamenetsky, M.D.: The Main Molecule. Nauka, Moscow (1983)
12. Green, N., Stout, W., Taylor, D.: Biological Science. Cambridge university Press, Cambridge (1984)

13. Hadorn, E., Wehner, R.: Allgemeine Zoologie. Georg Thieme Verlag (1977)
14. Hopson, J.L., Wessels, N.K.: Essentials in Biology. McGraw-Hill Publishing Company (1990)
15. Capra, F.: The Web of Life. Anchor Books, Doubleday, New York (1996)
16. Gorbachev, V.V.: Concepts of Modern Science. Ed MGUP, Moscow (2000)
17. Medawar, P., Medawar, J.: The Life Science. Current Ideas of Biology. Wildwood House, London (1978)
18. Mayr, E.: Populations, Species and Evolution. The Belknap Press of Harvard University Press, Cambridge (1970)
19. Lewin, R.: RNA catalysis gives fresh perspective on the origin of life. Nature **319**, 545–546 (1986)
20. Hochachka, P., Somero, G.: Biochemical Adaptation. Princeton University Press, Princeton (1984)
21. Wilson, A.S.: The Molecular Basic of Evolution. Scientific American, pp. 164–173 (1985)
22. Odum, E.P.: Basic of Ecology. Saunders, Philadelphia (1983)

# Part II
# Living Bodies: Carriers of Life

# Chapter 7
# A Mode and a Tool of Life

## 7.1 The Nature of Living Bodies

The notion of "life" is always associated with the presence of specific bodies. This is certainly true, since life cannot exist by itself without a carrier. Such carriers exist as various forms of bodies of many different types of organisms.

A living being is not an abstract notion, but rather a concrete one. It is a material body, a rudimentary unit of biological activity, and a carrier of life. Every organism has its own peculiar characteristics, leading to the concept of individuality. Each individual organism is a basic complex organizational unit of a species, which possesses the characteristics of living matter. Individuals are representatives of various species, and are therefore specific organisms that possess the peculiarities of an allelic composition of a genome, and as a result, some peculiarities of a phenome. All individuals are genetically programmed to express behaviors that secure their survival because it is important to save every unit of life that contains a precious genome. In other words, the strategy of life of any individual is survival, reproduction, and dissemination of genomes.

Organisms are very different in their natures, forms, and sizes. Living bodies can be unicellular, multicellular, or noncellular (spores and viruses). Their sizes can vary from 20 nm (HIV virus) to 30 m (blue whale) and more.

The diversity of organisms is studied by the science known as ecomorphology, which creates complex classifications similar to those established by modern taxonomy. At every level of organization, ecomorphology classifies numerous specific body forms typical for living organisms. Multicellular beings are classified as having monomeric or metameric bodies, whose form is determined by either the prevalence or lack of locomotion (Fig. 7.1). These traits have arisen as a result of long evolutionary processes and adaptations to specific conditions of existence as well as interactions with other species. However, these are all living bodies—organisms that possess common properties and characteristics. Their internal components are all separated from the external environment by various coverings, and they are autonomous and possess both a genotype and a phenotype. They are

G. Zhegunov, *The Dual Nature of Life*, The Frontiers Collection,
DOI: 10.1007/978-3-642-30394-4_7, © Springer-Verlag Berlin Heidelberg 2012

**Fig. 7.1** Various forms of living bodies. In spite of significant differences, all these organisms are integrated by the presence of the genome and by cellular organization. They are all the products of the activity of cells under the management of their individually specific genomes. Organisms often contain non-cellular substances, such as the inner and outer skeletons of many animals and the bark of trees

also capable of reproduction and possess the ability to maintain their specific organizations over millions of years and throughout trillions of generations.

Organisms are a part of Nature and exist in unity with the environment. As mentioned previously, they are characterized by specific interactions with the environment in the form of matter and energy exchanges. These are interconnected processes, since together with the flow of organic matter, chemical energy can be transferred in order to create order and maintain homeostasis. Living bodies carry huge amounts of external substances and energy throughout themselves. For instance, an average sized adult who consumes 3 kg of food per day for an average of 70 years utilizes approximately 70,000 kg of external substances during his or her lifespan!

Living bodies are characterized by high levels of structural and functional order. Organisms develop, function, and maintain order due to a multitude of molecular and cytogenic processes that take place every second in many trillions of cells. From this point of view, living bodies are the physical outcomes of the combination of all "predetermined and structured processes". Highly structured order comes from the utilization of genetic-informational mechanisms which serve as the blueprint to order the movement of various types of matter at different organizational levels within living bodies. When considering such bodies, structures may be thought of as being synonymous with processes, since the various permanent structures within a body are essentially the results of prolonged existences of molecular systems and their associated processes where internal connections are stronger than external forces. In addition, there exist molecular processes of metabolism, where short-term interactions and internal connections are weaker than external forces. Hence, it is possible to say that living bodies are a totality of processes targeting the organized movements of matter.

A wide variety of every possible phenotype of living bodies is conditioned by the presence of the same large number of genomic options. The realization of different aspects of genetic information during the process of development leads to the formation of various bodies. Living organisms differ in principle from non-living materials by the fact that they possess not only special structures and functions—(phenotype), but also programs for organization and manufacture of similar new organisms (genotype). These programs can replicate repeatedly (DNA replications) and can be passed from one generation to another during the process of propagation. Even the most complex program-controlled robots created by men, as well as computers and their systems that carry out functions beyond the abilities of a human being, still do not include programs for their own reproduction and development.

Living bodies have a definite life term. In the time frame of their existence, they gradually wear out, get old, and die. However, a majority of individuals leave descendants that provide for the genetic continuity of life. Essentially, the main purpose of bodies includes: (a) the creation of the conditions needed to form germ cells; (b) direct production of gametes that contain a permanent genome; (c) transfer of a genome to subsequent generations. Gamete bodies are a transitory form of life's existence, because their sole purpose is to carry an exact single copy of a specific organism's genome, which will be used to create similar new organisms during the process of syngenesis. This particular form of life is similar in structure and mechanism across the majority of all organisms, and is essentially the existence of a genotype without a truly realized phenotype. Hence, it is important to emphasize that a living organism is one of the stages of the continuous life of a genome, a process that consists of alternate forms of existence of gametes and their producers. In other words, living organisms can be considered as one of the stages in the development of a genome. The phenotypic framework of even one genome may have various manifestations, which is called polymorphism. It is through this characteristic that various differentiated cells, tissues, organs, as well as different and unique organisms of the same species, are established. Larval and immature forms of insects that have a single genome, for example, look totally different from the adult organism. The idea that living beings are just "machines for survival of genes" was first stated by R. Dawkins in 1976.

Thus, all manifestations of life are connected with the existence of highly organized, integral, and hierarchic systems of molecules and cells, which are formed and supported by the genetic programs within specific genomes at all stages of life. Living bodies are "hubs of realized genetic information" and are built from it, live by it, and survive by it. In living organisms, every point of organization corresponds to certain flows of information, starting at the molecular level. At this level appear systems of macromolecules (NA) with properties of self-reproduction and maintenance of integrity. An "informational explosion" from a complex of these molecules (Fig. 4.1) creates a cascade of consecutively managed transformations of matter, energy, and information that finally cause the formation of a specific living body in a particular space and time. Such an informational explosion takes place, for example, right after the fertilization of an ovum.

The phenomenon of life, therefore, is connected with the existence of highly organized bodies that co-exist and interact with the environment. Living organisms are discrete units of phenotypic life that represent temporary self-reproducing systems, the reason and basis of which is a permanent genome. In other words, living organisms are not a goal, but only a means; they are not the reason of life, but just its consequence.

## 7.2 Movement and Activity

Living bodies are complex systems whose existence is based on the global ability of movement of all their components. To move means to possess energy. Practically all forms of movement have significant meanings for living bodies, since they provide for a multitude of interactions, as well as metabolic, physiological, and behavioral processes. With respect to living organisms, various motile processes need to be considered at different levels:

*The movement of an organism as a single unit*. For individuals, it is important to look for food and favorable living conditions, to avoid unfavorable conditions, and/or to search for a sexual partner for propagation. The "behavior" of various organisms is rooted in conditioned, expedient, controlled, and purposeful movements. Different organisms have different behaviors due to variations in reflexes and instincts, which are defined by genetic programs characteristic of a specific organism's genome. It may be said that the genome is the "music" that makes living bodies "dance."

*Movement of cells*. Multicellular organisms contain freely living cells, most of which do not have a permanent location and are capable of active or passive movements. For example, erythrocytes, thrombocytes, and leukocytes move passively with the blood flow. Lymphocytes, macrophages, and neuroglia cells are capable of active and independent amoeba-like motility. The dislocation of cells is also very important at early stages of embryogenesis. Some cells from different parts of an embryo can migrate to other places and form a specific population which may be used later for the formation of organs.

*Movement of cytoplasmic components*. All components of the cytoplasm are in constant motion. Every molecule in a biosystem carries out various types of thermal motion such as linear motion and rotary motion, as well as individual oscillations. All biochemical reactions are the end-products of dynamic molecular and atomic interactions which are mediated by movement. Brownian motion, a particular kind of random movement in a medium, is the basis of life, since it serves as the general type of motion experienced by cellular components, and therefore provides for the contact and interaction of molecules leading to all the biochemical processes of metabolism (Fig. 7.2).

Therefore, ordered motion at all organizational levels of living organisms is one of the main conditions for the flow of vital processes which contribute to their existence and survival. The orderliness of bodily structures is mainly required for

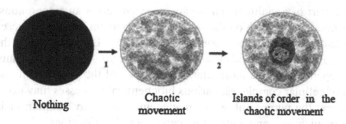

Nothing      Chaotic movement      Islands of order in the chaotic movement

**Fig. 7.2** Movement and life are inseparable. 1—Some powerful incentive causes chaotic thermal motion and matter to arise simultaneously. 2—Afterwards, an informational incentive singles out a particular form of motion.Without order, molecules, atoms, and organelles would undergo chaotic and random movements, which would never result in the creation of life. The informational program of the genome organizes the motions, establishing a separate peculiar region within the chaos, leading to the creation of life. Thus, life comes into existence on the basis of ordered thermal motions

the isolation and organization of processes which act against naturally disintegrating and destructive forces of various internal and external media. In other words, on the basis of all the different types of movements, a constant activity occurs which allows for the application of continuous work to be directed against unfavorable factors that tend to increase entropy and unbalance the living system.

## 7.3 Autonomy and Survival

Living bodies possess an aggregate of properties that differentiate them from non-living bodies: nutrition, respiration, excretion, movement, reproduction, etc. However, no single specific property can be named which alone defines a living organism. It is a complex of such qualities that determines this qualitatively new condition of material existence.

One of the main characteristics of a living body is its autonomy. Living organisms are definitively segregated from their environments. Each organism has a body which has a specific size and shape, as well as the ability to protect its internal contents. These features provide organisms with a relative independence from the environment and aid in maintaining a constant internal medium.

Living systems have developed a multitude of processes that are not characteristic for non-living matter. These processes are unique and take place only under conditions of stable internal cellular contents. Metabolic, physiological, and other vital processes are clearly managed by a genetic-cellular system, and are also regulated by other cells within the organism. The functions and behaviors of biological bodies, which arise as a result of the unique and specific presence and regulation of such processes, are therefore directed mostly by internal systems. These bodies move and work not just under the impact of external forces, but primarily on the basis of internal reactions to external actions or by carrying out internally driven motives.

Even as a part of a multicellular organism, each cell is an autonomous system which does not depend much on the external environment due to the presence of a membrane and the regulated support of homeostasis. By maintaining their own molecular equilibrium, cells also control the homeostasis of the organism they comprise. Depending on the molecular composition of the cells and the degree of orderliness of cellular complexes, various biochemical processes may occur which primarily target the purposeful use of matter and energy to maintain their autonomy and organization, and only then carry out the functions of an organ or organism.

Multicellular organisms are built and function as autonomous single units due to the integration of all the molecules, cells, tissues, and organs, as well as their coordination by various systems. Thus, both cells and multicellular organisms are self-regulating systems that are driven by genetic programming in order to support and maintain their structures and functions at each organizational level.

One of the main goals of all organisms is survival. Most organisms live in a rather aggressive environment. For example, although most land organisms are composed of up to 75 % water, they live in a gaseous environment (air). Different organisms can live in frozen tundras, dry deserts, hot springs, and so on. Despite having different and sophisticated methods of adaptation, no single creature has the ability to survive in all environmental conditions, which means that all creatures must constantly fight for their lives.

In order to survive, individuals must first protect themselves from unfavorable factors of the external environment (cold, heat, drought, radiation, etc.) or attempt to avoid them. It is also important to elude adversaries, e.g., to evade predators. Under such extreme conditions, weak and defective representatives are "weeded out" through the mechanism of natural selection.

Living organisms need to constantly procure nutrients in the various forms required to drive different metabolic processes, allow for the constant regeneration and expansion of their molecular composition, promote increased cell division, and support the organism's integrity. Nutrition also provides organic substances for oxidative processes and energy storage and utilization through various forms such as ATP. Nutrients are generally far from being accessible at all times and in sufficient quantities. Therefore, in order to survive, many organisms need to endure tough competition and it is mainly the strongest representatives that survive or remain intact long enough to try again. From the dual points of view of the essence of life as the "existence of evolving genomes" as well as the essence of complex organisms as multicellular colonies, it is those colonies with the best and most adaptable genomes within specific living conditions that are most apt to fight and survive in those environments.

Every organism is programmed by Nature to reproduce. When not fighting for survival a majority of time is spent either in search of a mate or preparing for processes of reproduction. This provides for the maximum number of individuals, and therefore the long-term survival and existence of the species. Even this highly vital process is subject to competition, which generally leads to the selection of the best candidates to produce the heartiest offspring in order to propagate and protect the species. In this most important process of reproduction, there also exists competition that leads to the selection of the heartiest genomes. Strong parental genomes condition the appearance of strong phenomes in posterity.

The average duration of the existence of a species of living organisms on Earth is somewhere in the range of several million years. The average life-span of individuals can vary significantly, ranging from 20 min (bacteria Escherichia coli) to several hours (mayfly Ephemera vulgata) to thousands of years (pine Pinus longaeva), but it is still significantly shorter than the existence of an entire species. The basis for the prolonged existence of a species is the ability of individuals to transfer the hereditary information from one generation to another through reproduction. Because of the inherent ability of genetic information to replicate and modify, new individuals can be constantly observed, some with new characteristics that may be beneficial in acclimating the new organisms to their environmental niches. This strategy allows species to survive and exist during very long periods and also to gradually evolve, propagate, and adapt to new conditions.

Thus, the biological basis for living individuals can be expressed by two aims: survival and propagation. This, in turn, secures the preservation and propagation of specific genomes, which maintain the stability of the existence and development of the Integrated Global Genome. For higher functioning organisms such as human beings, the strategy of life is also expanded to contain various personal and social goals such as education, wealth, career, status, and so on.

# Chapter 8
# Cells and Organisms

## 8.1 Unicellular and Multicellular Organisms

We may define both individual cells and multicellular bodies as organisms, because we define an organism as a united, complex, and organized manifestation of life. Although both cells and multicellular bodies embody this definition of an organism, these systems are fundamentally different. Cells are heterogeneous systems of millions of different molecules and their complexes, while multicellular organisms are systems that consist of such cells, which are largely similar and organized into macro-structural units. In multicellular organisms, entire blocks of cells form autonomic structures, as well as functional complexes of significantly large sizes. In other words, multicellular bodies are globally simpler in composition than their single constituents, while on the other hand, they are also more physiologically intricate since they unite several levels of increasing complexity. This is why it is not always possible to extend principles such as the construction, function, and behavior of cells to multicellular organisms, and vice versa. Nevertheless, it should be noted that certain properties such as sensing and adapting to the environment and maintaining integrity, motility, and communication are shared characteristics between single cells and the organisms they comprise, just at very different scales and through different mechanisms.

There are also important biological structures that do not have a complete cellular construction, although they are either derivatives of cells or parts of their life cycle. For example, erythrocytes are unique formations which do not have a complete cellular structure because mature cells do not contain any genetic material, mitochondria, or many other cellular elements. Their cytoplasm contains mostly hemoglobin molecules with a minimum number of enzymes. Because these cells cannot propagate, it would be difficult to claim that erythrocytes are real cells or true living bodies. Nevertheless, at earlier stages of maturation (erythroblasts), these cellular constructions have all the same components as "living cells". Another example of such biological duality can be seen in the trophoblast cells of the developing embryo, which are "alive" until later developmental stages when

G. Zhegunov, *The Dual Nature of Life*, The Frontiers Collection,
DOI: 10.1007/978-3-642-30394-4_8, © Springer-Verlag Berlin Heidelberg 2012

they become inert and form a large part of the placenta. These observations imply that not all cells are necessarily required to remain "alive" to continue carrying out their respective biological roles.

The emergence of multicellular organisms has been one of the most important stages of the evolutionary process. From a combination of different individual cells, it became possible to create numerous varieties of multicellular organisms with differing properties and functions and with new principles of interaction with the environment. This has significantly increased the abilities of cells and their genomes to adapt and propagate by utilizing the systems of the entire organism. The discrete organization of a being from trillions of self-replicating cells assists the gradual replacement of dying cells or pathologically altered body parts without disturbing the vital activity of organs or the organism as a whole. This significantly increases the lifetime of a multicellular body and its genome. Due to constant cellular and inter-cellular regenerations, there is a molecular control of all aspects of large multicellular organisms, because only the constituent cells possess the capability to keep, repro-duce, transfer, and realize genetic information. The organization of a multicellular organism from small morphological units (cells) that possess large surfaces is very favorable for the metabolism and energy exchanges with the environment and with each other. The distribution of functions between the cells in a multicellular organism provides wide opportunities for development and adaptation. By maintaining their integrity, cells control the homeostasis of the organism they are a part of.

Based on the idea that a cell is first and foremost a receptacle for a genome, multicellular organisms can be imagined as colonies of cells that incubate discrete genomes. These cells are united for mutually beneficial existence, survival, and reproduction. Such organisms are rightly considered to be the most complex communities of the smallest living bodies. This is a qualitatively new stage of existence of cells and their genomes. Concurrently, all the differentiated cells act in concordance and are subject to the demands of their genetic apparatus and the regulatory systems of the organism. They maintain the integrity of the organism and provide for a variety of its functions.

Since cells and multicellular organisms are completely different biological systems, their principles of reproduction, development, integrity maintenance, ageing, and death are also significantly different. Single cells propagate mostly by division, have a generally simple developmental program, and use molecular mechanisms to maintain integrity. Multicellular organisms, on the other hand, propagate by gamete formation and embryogenesis, and have highly complex developmental processes with many stages. Furthermore, they utilize various cyto-logical and physiological processes of regulation and regeneration to maintain integrity, and their ageing and death are connected with the disruption of manage-ment, control, and coordination of cells, tissues, organs, and systems. Multicellular organisms also possess an aggregate of new properties. In particular, they have more complex systems of feeding, digestion, respiration, excretion, reproduction, move-ment, homeostasis maintenance, growth and development, adaptation, etc.

It is important to note that the genome of a multicellular organism contains two significantly different programs. The first one is connected with the formation and maintenance of integrity and metabolic activity of cells. It is associated with mechanisms such as replication, transcription, translation, cell division, regeneration, homeostasis, and so on. This program is rather clearly defined and is based on the principles of propagating and realizing genetic information (DNA $\rightarrow$ RNA $\rightarrow$ protein). The second one is also connected with processes such as the formation and maintenance of integrity, but with respect to the functional activity of macroscopic cell unions. These include the programs of gametogenesis, fertilization, individual development, maintenance of a histological and anatomical structure, various functions, etc. Unlike the first program, this one is not as clearly defined. For example, it is not clear how information about functions, forms, sizes, and localization is recorded and saved in the genomic memory. There obviously exists a tight association between the first and second programs, since the functions of the organism are determined by the combined work of every cell, and the work of all the cells is in turn controlled by various stimulatory and regulatory systems of the organism.

Multicellular organisms are gradually and meticulously created from a single mother cell called the zygote, which results from the fusion of two parental gametes. Cytologically speaking, the ability to create multicellular bodies has much to do with the ability of cells to intensively divide, differentiate, adhere, and migrate. Adhesion is the ability of cells of one clone to connect to one another or bind to extracellular surfaces, and it is a property that varies depending on the cell type. At earlier stages of embryogenesis, cells actively travel, recognize each other, and accumulate. This leads to the formation of groups and layers of cells which later differentiate by selective expression and communication, leading to the creation of tissues, organs, and, finally, the organism itself.

We have presented clear evidence that multicellular organisms differ significantly from unicellular ones. Such organisms represent a relatively new level of development in the phenomenon of life, and possess unique and infinitely more complex physical and abstract features, such as the brain and consciousness. Unicellular organisms (like Monera and Protista) exist autonomously and almost independently from each other, while cells of various multicellular creatures are tightly connected and exist with mutual beneficence. There are many options for the consolidation of cells into multicellular organisms, the properties of which differ significantly from the properties of discrete cellular elements.

In this section, a duality of the organization of living bodies can be sensed. On the one hand, it can be an autonomous egoistical cell, the basic structure for a carrier of life, and on the other hand, it can be an independent and autonomous multicellular organism.

## 8.2  Inside the Living Cell

All cells possess a complex internal organization that provides synchronous concordance of hundreds of highly complex processes in various parts of their "bodies". Mammals have over 200 types of cells that are specialized in the realization of specific functions, and possess various peculiar features and structures in order to carry out their selected roles. Nevertheless, all cells have a standard set of internal differentiated "body" parts, without which their "personal" and "public" activities would be impossible to perform, and include various organelles, membranes, cytosol, cytoskeleton, karyoplasm, etc. Specialized cells have certain specific properties with respect to their composition and organization, which depend on their roles. For example, animal muscle cells are tightly packed with myofibrils, have an elongated form, and possess specialized organelles such as the sarcoplasmic reticulum, plasmalemma, and so on. Neurons have specialized outgrowths called axons and dendrites that function in cellular communication. Secretory cells are full of various other elements such as the endoplasmic reticulum, Golgi apparatus, secretory vesicles, etc. The cells of various other kingdoms of living organisms also have their unique peculiarities. For instance, plant cells have specific compartments such as chloroplasts, vacuoles, and cell walls. Prokaryotes, on the other hand, do not have nuclei, nor a majority of the other aforementioned organelles and features. It is also important to note the high level of organization of the intercellular fluid which is specifically formed by each different cell type.

Thus, cells are extremely complex dynamic systems. Such systems consist of ordered macromolecules which form membranes, compartments, and organelles of various sizes, complexities, and assignments. No single component of a cell can exist without interacting with other components, much as a multicellular organism cannot exist without coherent cross-talk between all its cellular constituents. It is for this reason that life is a property of the entire cellular system (genome + phenome), be it a single cell or a complex multicellular being.

Intercellular organization is characterized by high negentropy (syntropy), and therefore a high degree of informational value. A significant portion of this information is transported non-genetically to later generations during the division of cells. The main component of the cells is a nucleus, where the genetic apparatus is located and where the main molecular and informational processes of management of other cellular parts take place. The mechanism that controls and maintains order is connected with a differential expression of genes and the synthesis of special proteins and enzymes, as well as through the activities of distinct cellular compartments. Mitochondria, for example, are structures that provide, among other things, a steady supply of ATP, the cellular energy currency, for power-driven events. Various cytosolic enzymes are informationally interconnected and selectively catalyze thousands of biochemical reactions, leading to the coordinated maintenance of metabolic activities and a variety of other vital functions. Furthermore, cells possess structure-functional flexibility, which manifests itself in the cooperative functions of various organelles and proteins that work together to achieve different goals. Thus, powerful functional systems are formed in order to perform more important global functions attributed to the whole body. For example, in the processes of energy transformation in cells, many systems such as membrane transport, cytosolic transfer, and internal digestion work in unison with mitochondrial metabolic reactive complexes and hundreds of enzymes to generate ATP.

In principle, the main content of a cell is the same regardless of the type of organism it is derived from, even though specialized adaptations can be quite different. These main parts are the genome and the cytoplasm which surrounds it. With reference to our previous simile of cells as biological computers, we may say that the whole complexly organized colloidal content is a system unit that provides quick and precisely targeted flows of matter and energy. The genome acts as the "processor" which controls and regulates these flows via the genetic "operating system". Together, they create a nanoscale cybernetic operation-executing biological system called a cell.

The mechanisms of cellular organizations within living bodies are still quite poorly understood. We know that a multicellular organism is a complex ordered system of relatively independent cells. We also understand the reasons for their consolidation and interrelated functions because various systems of control and regulation of homeostasis have been quite well investigated in different organisms. It is also quite clear that molecules in cells behave according to the laws of physics and chemistry. Their behavior is compliant with the laws of solubility, chemical kinetics, thermodynamics, electrostatics, etc. Nevertheless, the underlying mechanics in the cells, which utilize these laws with such precision, are infinitely more complex. For example, it is still not well understood which necessities initiate the division of cells, how targeted transportations of substrates and metabolites are managed, how it is possible to carry substrate molecules from one enzyme to another, and what manages organized movements of giant intracellular masses such as DNA during mitosis. It is also not clear how it is possible to build, for example, a complex molecular machine like a working ribosome. How does it function as a single unit and gracefully interlace into the organized network of other structures and functions of the cell? What are the conditions for the expediency of its activities? How do isolated ribosomes and other organelles work outside the cell and maintain their properties and characteristics for prolonged periods? Because they appear so independent of the cells, does it mean that they are relatively autonomous living bodies themselves? These are just some of the myriad of questions relating to cellular events and structures. Moreover, the nucleus, which has the highest degree of complexity in its composition, raises even more questions. How does it work so clearly and expediently, carrying out the most complex molecular processes with such precision? What biological laws maintain this most complex apparatus which appears to be able to sense its cytoplasmic surroundings in order to control them? It appears to exist as a structure that has its own life, with rather autonomous and poorly understood behavior.

To attempt to resolve any of the above quandaries, we must present the element of the cellular system which has the ability to control everything and act in a targeted way. This element is the genome, which controls all the structures and functions of both individual cells and the multicellular organism as a whole. Moreover, the expediency of its actions is determined by the tasks of self-preservation, survival, and propagation. However, what manages the genome itself? This is one question which is still far from being answered.

## 8.3 Biological Membranes and Compartmentalization

*Biomembranes*: The various structures and functions of cells, organelles, tissues, and organs are all dependent upon and maintained by the presence of biological membranes. Such membranes are one of the main elements of cellular organization, and can be intra or peri-cellular.

A fluidic bilayer sheet (the plasma membrane) covers the outside of all cells with a very thin film (about 5 nm thick) which is composed of two "leaves" of amphipathic lipid monomers (polar on one side and nonpolar on the other). This membrane separates the cells from their environment and allows them to maintain a specific internal molecular composition. Without membranes, the contents of cells would just spread out, and this would cause a loss of order of elements of the cellular system and an unfavorable thermodynamic equilibrium that would mean certain death. In eukaryotic cells, membranes also cover internal organelles and form a branched intercellular network.

The lipid and protein composition, as well as the architecture, of membranes can vary significantly between different cell types and organisms, and even the membranes of different organelles within a single cell can have structural peculiarities. Nevertheless, the membranes of most organisms are based on the same basic organizational principles. First and foremost, they are all formed from a double layer of phospholipids with embedded and peripheral proteins. The fatty-acid tails of the phospholipids, as well as water-insoluble domains of various proteins, form a hydrophobic inner core while the polar head groups and charged protein domains form the extracellular and intracellular sheets. This layer acts as an impermeable barrier for practically all water-soluble materials, which generally require specialized protein channels and transporters for cellular entry and exit. Thus, the membrane is the totality of interacting molecules that determines a new systemic feature: the selective support of the functions of the genome's phenotypic framework.

Proteins comprise approximately half of the composition of the plasma membrane. These can include molecular carriers, transporters, enzymes, and receptors to various hormones or other stimuli. The double layer of phospholipids has a unique liquid crystal nature, where specific lipids and proteins can move in various lateral directions, and sometimes even switch between leaflets. Thus, membranes are highly mobile and elastic molecular systems which carry out or participate in the execution of many different roles. The lipid and protein compositions of various membranes found in different cells and organelles condition their unique and distinct functions. Hence, biological membranes are one of the most important structural and functional elements of complex living systems, providing for their many properties. The emergence of functioning membranes and the separation of protobionts from the environment was one of the key steps towards establishing the full autonomy of living bodies and improving their interaction with the environment.

*Compartmentalization*: This is the patterned division of individual cells and multicellular organisms by membranes into distinct, functional micro-environments (blocks), which allows them to simultaneously carry out many thousands of biochemical reactions (often oppositely directed) and perform a multitude of different functions in a coordinated yet independent manner. Each part possesses its own functions, yet exists and works in close connection and interaction with the others, thereby allowing the organism to perform multiple complex tasks. For example, many structural and functional body parts participate in the process of thermal regulation in mammals. In particular, these include temperature receptors, the hypothalamus, thyroid and adrenal glands, muscles, adipose layers, skin, the circulatory system, liver cells, mitochondria, etc. The advantage of such a "block principle" is that starting with single cells and working up to the entire body, organisms can use different combinations of structural and functional "blocks" in different situations, forming dynamic functional systems to carry out or maintain various functions.

About half of the actual volume of a eukaryotic cell is divided by membranes into sections, many of which are known as organelles. These can include, among others, the nucleus, cytosol, endoplasmic reticulum, Golgi apparatus, mitochondria, lysosomes, peroxisomes and chloroplasts. Each block contains specific enzymes that selectively catalyze only specific biochemical reactions which define its unique functions, a decisive factor in the separation and ordering of the many and varied processes that concurrently take place in a coordinated manner in the cells. For example, the cell nucleus contains the major portion of the genetic material (5 % is contained in mitochondria) and is the primary location for the synthesis of DNA and RNA. The cytoplasm, which surrounds the nucleus, consists of the cytosol and various organelles mentioned above. The cytosol is also a compartment that comprises slightly more than half the total volume of the cell, and serves as the principal site of protein synthesis and the majority of metabolic reactions which provide for the necessary construction and function of the cells and their respective "blocks".

Much like single cells, multicellular organisms are also built on the block principle. One particular and voluminous compartment in most animals, for example, is the extracellular space, which is composed of both vascular and interstitial spaces. In humans, for example, this space may contain liquid volumes up to several liters. Many specific metabolic processes take place there, along with the accumulation and modification of many nutritive materials. Oxygen, carbonic gas, amino acids, glucose, and hormones are just some of the many substances that transit through it. The circulatory system, which is responsible for blood circulation and is mainly composed of the cardiovascular system, is closed, meaning that it is separated from the cells of the organism by vesicular walls, and is also a section of the organism where specific contents, including many of those that are found in the extracellular space, are trafficked throughout the body in order to perform various tasks. Every organ can be considered a specific body "block", i.e., a separate part of the body that has a specific size, form, location, structure, and special functions. However, despite the structural separation of the parts of a cell or an organism, they still manage to function as a single coherent unit due to the presence of special means of communication via signal processing between individual parts of cells, between cells, and between whole body parts.

Lastly, the block principle of organization can also be used to refer to various processes and mechanisms, as well as functional molecules created by Nature. For example, there are special protein complexes, such as ribosomes and G proteins, which are composed of multiple subunit "blocks" that come together to assemble into functional units. Each element and reaction in a vital process such as DNA replication can also be thought of as a block, which works together with other such blocks to achieve the desired biochemical role.

One of the main merits of compartmentalization is the possibility of progressive evolutionary events that can lead to the creation of new biosystems or biochemical processes based on the combination of various already known and tested structural and functional blocks. One example of such a process is the mechanism of aerobic breathing, which formed in the later stages of life's development on the basis of earlier anaerobic processes. As already mentioned, because many protein complexes are composed of multiple domains, there is a chance that structural changes or incorporation of new subunits driven by evolutionary pressures may potentially establish new biochemical pathways. This may be achieved, for example, from the directed or spontaneous recombinations of genomic nucleotide "blocks", which can lead to the creation of new proteins. Even the establishment of eukaryotes resulted from the union of structural and functional blocks of archaic cells. Depending on the combinations of various blocks and their further developments, new characteristics such as autotrophy and heterotrophy eventually appeared. The formation of colonies from these new "eukaryotic" cells resulted in a new leap in the evolutionary cascade, which eventually led to the emergence and development of multicellular organisms.

Thus, in summary, the principle of compartmentalized unit organization is common to the structure, function, and evolution of all living systems, from single cells to entire organisms. It allows biological systems to react quickly and precisely to various environmental changes and to maintain homeostasis and aid in survival. These roles are achieved, first and foremost, by the coordinated interactions of all the standard parts of living systems, as well as by the reversible dynamics of structural and functional mechanisms which are formed from these various "blocks".

# Chapter 9
# Reproduction and Individual Development

## 9.1 Reproduction

Reproduction is one of the main properties of living bodies, and is an obligatory condition for organismal life and for the continuous existence of different species. Species propagation is a constant process of reproduction, creating new generations of representatives from all types of organisms. The existence of each specific cell or multicellular organism is temporally limited, which is why the "immortality" of a species can only be achieved by reproduction. All species consist of individual representatives, each one of which eventually dies, while, due to the phenomenon of reproduction, the life of a species continues.

There are two main types of reproduction which will be discussed here—asexual, which takes place without the participation of gametes (without exchange of genetic information), and sexual, which is characterized by the formation of gametes, fertilization (with exchange of genetic information), and formation of embryos. Although both will be covered, we will focus more on sexual reproduction, especially as it pertains to animals.

*Asexual reproduction has the following characteristics*: (a) only one parent is involved in the process of reproduction; (b) no formation or merging of gametes takes place; (c) in the basis of reproduction, there is a replication of DNA and mitotic distribution of the genetic material; (d) new specimens can develop from a somatic part of the parental organism; (e) daughter organisms are genetically similar to each other and the parent; (f) the process provides for a quick increase in the number of specimens; (g) the unit of reproduction can be either the whole parent body, or a part of the body, or a single somatic cell.

Asexual reproduction is mainly typical for prokaryotes and single-celled organisms, and for many fungi and plants. It has a great value for animals as well, since the formation of multicellular organisms occurs through sequential mitotic divisions of each cell. Cell division ensures the growth of an organism and the regeneration of its tissues and organs.

G. Zhegunov, *The Dual Nature of Life*, The Frontiers Collection,
DOI: 10.1007/978-3-642-30394-4_9, © Springer-Verlag Berlin Heidelberg 2012

The distinct value of asexual reproduction can be seen in the rapid and effective elevation of the number of individuals which carry a particular genome. In a constantly changing environment, asexual reproduction ensures a prompt spreading of successful combinations of genes. Therefore, organisms that primarily undergo asexual reproduction (bacteria, numerous protozoa, and fungi) possess colossal progressions of growth.

*Sexual reproduction has the following characteristics*: (a) usually two individuals participate in sexual reproduction: a male and a female (with the exception of hermaphrodites, which in certain cases are capable of self-fertilization); (b) the process is characterized by the formation of gametes (gametogenesis) and their merging (fertilization); (c) one of the stages of gametogenesis includes meiosis, during which transformation of genetic information occurs; (d) sexual reproduction is characterized by a high genetic variability where daughter organisms differ from their parents and from each other by the allelic composition of their DNA; (e) the units of sexual reproduction are haploid gametes; (f) the speed of sexual reproduction is rather slow due to the time needed to form gametes, search for a partner, fertilize, and develop a new organism.

Sexual reproduction consists of several highly organized molecular, cellular, and genetic processes:

- Copying the program of development—replication of DNA molecules;
- Archiving genetic programs—chromosome formation;
- Exchanges of DNA segments between homologous chromosomes by crossing-over;
- Transfer of the genetic programs to daughter cells—mitosis or meiosis;
- Formation of the transitory form of genomic existence via gamete formation;
- Union of male and female genomes by fertilization and zygote formation;
- Realization of the genetic program of development on a cell-by-cell basis.

It is important to note that all the above processes are connected with operations performed by and on the hereditary genetic material. In other words, the basal structure of sexual propagation is the genome, and the essence of the process of sexual reproduction lies in molecular manipulations of DNA and its exchange between partners in numerous generations.

The main vegetative phase of primal creatures consisted of a prolonged haploid genomic period. It has been suggested that during seasonal changes or unfavorable conditions, vegetative bodies began forming haploid gametes such as spores. After the fusion of two spores, a temporary zygote was formed that had a diploid genome. During adverse conditions, it became much easier for these organisms to survive. Upon the restoration of favorable conditions, new vegetative haploid organisms were formed from the zygote by means of further meiotic divisions.

During the evolutionary process, many multicellular organisms changed their priority concerning genetic ploidy. The duration of the diploid phase of genomic existence increased, resulting in their transformation into autonomic multicellular diploid organisms. At the same time, the main haploid phase of the genome was reduced and allocated primarily for reproductive events which included the formation of spermatozoa and eggs.

We should note that, at the moment of fertilization, life does not appear again as a new phenomenon, but rather as a transfer of "life substrates" (genomes) from one organism to another, thus forming new living bodies. Likewise, the organization of a cell does not appear anew, as it is already present in an ovule, and is maintained in the developing zygote. Then, through multiple cloning of a genome and via the orderliness of the processes of cellular division, a new complex multicellular organism appears that carries its genome and is ready to pass it on to further generations.

*The meaning of sexual reproduction*: (a) the appearance of sexual reproduction was a significant stage in the evolution of life on Earth, and is one of the reasons for the emergence of a multitude of living organisms; (b) through sexual reproduction, genomes from two different individuals of a specific species are mixed together, and this brings about genetic and phenotypic diversity. The genetic role of the parents in the determination of the features of the offspring is practically equal (same contribution to formation of genomes), in spite of the significant size and structural differences between an ovule and a spermatozoan. However, certain features are transferred only through the maternal ovule, including mitochondrial DNA, cytoplasmic organization, and factors for initial development; (c) sexual reproduction results in the generation of new gene varieties as parental genes combine and adapt; (d) sexual reproduction provides a population that is competitive in unpredictably changing environments due to the possible appearance of new features in some specimens; (e) in a larger population, sexual reproduction secures favorable alleles and removes unfavorable ones; (f) sexual reproduction maintains a diploid genetic composition. Diploid organisms possess an important advantage—they have alternative copies of each gene. These copies can mutate in order to serve as foundations in the creation of new features without leading to fatal consequences if a new characteristic proves inadequate to current environmental conditions. Diploidy also provides an organism with stability, since a harmful or lethal mutation of one of the gene copies is generally recessive, and therefore does not cause any appreciable harm. In many multicellular organisms, a diploid phase can be complex and prolonged, while the haploid one is simple and short-term. During the diploid phase, immediately after the fusion of gametes, cells of the developing embryo replicate and specialize, forming a complex organism; (g) in a majority of animals, it is possible to distinguish between germ line cells (primary gametes where new generations of gametes emerge), and somatic cells, which form the rest of the organism. Somatic cells mostly serve the needs of the germinal cells, especially supporting their survival, replication, and maturation; (h) sexual reproduction limits the accumulation of recessive lethal alleles, because if both heterozygous parents carry the same lethal mutation, their homozygous descendants will die out; (i) high combinatorial alteration provides the emergence and distribution of useful genes and features; (j) with periodical "rewriting" and "editing" of genetic information in each reproductive cycle, a prolonged stability of genetic individuality can be achieved within a given species.

**Fig. 9.1** Means of division
in unicellular organisms.
**a** Asymmetric division of an
amoeba. **b** Longitudinal
division of euglena.
**c** Transverse division of
infusoria. Regardless of
differences in their division
cycles, the main agent is
always a genome with
principal processes involving
multi-stage transformations
and simultaneous
transmissions of information

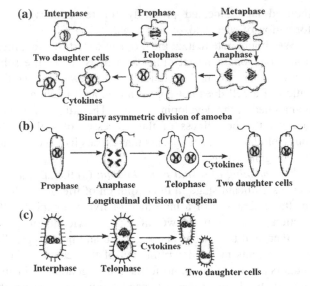

Many diverse means of sexual and asexual reproduction are known. Different types of animals, plants, and other organisms can have different mechanisms of reproduction depending on their anatomy and physiology, the environment, and their relative level of development. Nevertheless, the common bases for all types of propagation are molecular and genetic processes, especially the replication of DNA molecules, which are generally packaged into chromosomes (Fig. 9.1). Chromosomes are organized structures which contain the hereditary material of an organism combined with many proteins in order to form integral units that encompass all the information needed for the germination and development of new organisms. Replication causes the doubling of the hereditary material, which is later passed on to daughter cells in equal proportions through either mitosis (diploid cells) or meiosis (haploid cells). One may consider reproduction as a means of cloning and distributing DNA molecules in the Global Genome net.

In the process of fertilization, the genomic DNA of both parents, which is located in gametes, is transferred to new generations of individuals. The genomes of heterosexual organisms fuse when fertilized and form a zygote, the genotype of which is a new system of interacting genes that act as the foundation for the development of a new individual. Nucleic acids are the basis for regulating the synthesis of proteins. Through this synthesis, the genome regulates the reproduction and regeneration of practically all organic molecules within the cells, ensuring prolonged survival of the organisms. The central role of genomes and genetic mechanisms is therefore quite obvious in the processes of reproduction and development. In other words, the essence of any reproduction is a trans-organismal passage of genomes.

The reproduction of individual cells in multicellular organisms also plays a very important role. It is known that the majority of cells live for significantly shorter times than tissues, organs, or the organisms which they form. In order to ensure a prolonged functioning of these systems, a constant division and replacement of worn out cells with new ones is paramount. Furthermore, propagation on a cellular level causes the growth and differentiation of organs and tissues, such as those required for an organism to achieve sexual maturity, which is necessary for its own reproductive processes. The essence of the propagation of cells is division, i.e., the trans-cellular passage of genomes. It is important to note that constant transfer of the order of the cytoplasm and its components to new cells is important as well. It is through this process that information about the organization and functions of molecular systems within a cell is transferred via non-genetic means from generation to generation.

Reproduction conditions a number of very important properties in living organisms. Without it, important phenomena such as heredity, alteration, and evolution would not be possible. The ability of living organisms to transfer structural, biochemical, and physiological features to their descendants is called heredity. The mechanics of heredity are based on the ability of DNA to replicate itself exactly, and to transfer the hereditary information from one cell to another, and from a parental organism to its progeny. Alternation is a property that results in the emergence of different features due to genetic mutations and recombination. Together, heredity and alteration constitute the prerequisites of evolution and speciation.

The processes of the irreversible development of life, based on constantly changing generations of various organisms (or rather, generations of genomes), occur incessantly. Evolutionary processes can be very prolonged, and, currently, their most complex product is Homo sapiens, an animal that possesses true intelligence. Evolution is based on processes of progressive reproduction, hereditary alteration, competition, and natural selection. The action of these factors has resulted in the significant variety of life forms we know today, adapted to various types of environments.

*Gametes.* Organisms that reproduce sexually possess molecular and cellular mechanisms to preserve and package genomes into the tiny volume of gametes, which can then be saved, transported, and fused to form a zygote. Multicellular organisms have two significantly different types of cells:

1. *Somatic cells* form the whole variety of tissues, organs, and body parts. This class includes several hundred types of cells which vary depending on their functions and locations. Somatic cells constitute the majority of cells of an organism. They have a diploid set of chromosomes, which are divided and propagated through mitosis, ensuring that all the daughter cells are nearly genetically identical to the parental ones. Their main destiny is to ensure the survival of the organism, which in turn supports the production of gametes and

leads to genomic preservation. The idea of 'immortal inheritable plasma' was first suggested by August Weissman in 1914. He declared that the whole range of somatic cells of all organisms serves primarily to maintain the ability of an organism to store genetic material and further its propagation.

2. *Generative (germinal) cells* are formed in early embryogenesis. They then migrate to the gonads of an embryo, where they reside in the epithelium of sexual glands as immature gametes in the form of spermatogonia or oogonia. Upon maturation, these cells become true gametes, ready for their reproductive duties. Germinal cells can divide both by mitosis (the period of reproduction in gametogenesis) and by meiosis (the period of gamete maturation). Once mature, spermatozoa are incapable of division, whereas in many animals and other species, ovules can divide independently through parthenogenesis, or "virgin birth", thus forming a new living body without the need for paternal fertilization. Because embryogenesis often occurs in external or semi-external environments, gametes generally possess an ability to exist for some time, independently of the organism from which they came. They are therefore designed with all the provisions necessary to fuse their genomes under various conditions to form new organisms.

Gametes are unique cells, since only once they mature can they physically coordinate the processes of genomic and hereditary transfer from one generation to another during sexual reproduction, maintaining the genetic continuity of a species. Gametes are a result of various trends in the evolution and differentiation of cells of multicellular organisms. Eventually, a unique cellular line was formed which specialized only in the implementation of reproductive functions.

In comparison with somatic cells, gametes have a number of fundamental differences, the main one being that mature gametes have a haploid set of chromosomes that only contain a single copy of the genetic code of its producer, without its actual realization, until fusion with a partner. The genomes of male and female gametes contain equal hereditary information, so that after fertilization and zygote formation, diploidy is restored. For example, there are 23 chromosomes in human gametes, whereas a zygote and its future somatic cells will have 46 chromosomes.

Male and female gametes also differ significantly due to the need for each type of germ cell to carry out various functions. An ovule is a maternal gamete designed to be fertilized by spermatozoa, and then to develop into an adult organism of a given type. All the ovules of a mammal have a haploid set of autosomes and a single sex chromosome, the X-chromosome. It is the only cell from which a new organism can develop. Spermatozoa, on the other hand, are mature male gametes designed to transfer a male genome into an ovule. They possess the ability of active motion, which secures the meeting of gametes. The spermatozoa of mammals also contain a haploid set of autosomes as well as a single sex chromosome, which can be either X or Y.

Thus, the gametes of complex animals are highly specialized cells. In the process of evolution, they gained morpho-functional properties in order to carry

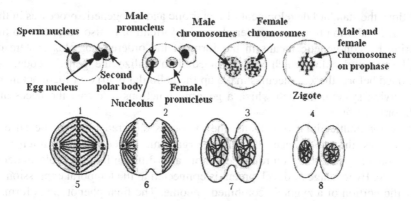

**Fig. 9.2** Cascade of transformation of the genetic material of mammalian gametes during fertilization. The phenome of an ovule assimilates a foreign genome and controls its amazing modifications. *1*—Penetration of a sperm into an ovary. *2*—Formation of a male and female pronucleus and DNA replication. *3*—Formation of chromosomes in pronucleus. *4*—Amphimixis—Integration of chromosomes from a mother and a father into a single qualitatively new system. *5*—Metaphase of the first division of a zygote. Chromosomes of a father and a mother are randomly aligned along the zygote equator. The system of division of chromosomes is formed. *6*—Anaphase. Each chromosome is divided into specific chromatids, which are distributed to opposite parts of the zygote. *7*—Telophase. Chromosomes unwind and form a network of chromatin, covered by a nuclear envelope. Two nuclei are formed and cytokinesis starts. *8*—Termination of cytokenesis and formation of a bi-cellular embryo that has its own genome from the combined genetic material of the mother and father

out the following special tasks: 1—To temporally maintain and conserve a set of permanent DNA molecules (genome) in a cellular micro-environment. During most of the life of the gamete, only the genome really exists while the phenotype is diminished. 2—To condition the union of parental genomes for the formation of a diploid zygote. 3—To induce the realization of the zygotic genome in order to develop a phenotype for the organism. 4—To finally guaranty the genetic continuity of life.

*Zygotes.* The zygote stage, which is a fertilized ovule, is a necessary manifestation in the development of all organisms that propagate sexually. It is the unicellular embryo of the future multicellular organism, and contains unified genetic hereditary material combined from the father and mother organisms (Fig. 9.2).

The ovule, followed by the zygote, plays an exceptionally significant role in the development of a new organism. It comprises many different regulatory molecules (messenger RNA, peptides, etc.) which are present in its cytoplasm, and without which the development of a new body would be impossible. For example, the cloning of organisms is possible only on the basis of a denucleated ovule. In other words, an ovule is a key structure in the appearance and development of new individuals. Such properties of ovules, and also zygotes, are stipulated not only by a genome, but also by the presence of a high inner order of molecular and supramolecular structures. This orderly organization is handed down to descendants and conditions the standard development of metabolic and cytogenetic processes in the progeny. In addition to genetic material, vital information is also passed on to the offspring by non-genetic means in the form of the ordered complex system of cytoplasmic molecules which will be needed to realize the genetic code. As mentioned before, these molecules make up the colloidal matrix, or the machinery of the living system, through which a genome functions to create the necessary conditions of life.

Thus, for multicellular organisms, the zygote is the cradle of life, because it alone allows the emergence of a new organism. Based on the differential expression of genes, the controlled division and differentiation of blastomeres takes place. Every stage of development is connected to the temporal expression of a specific portion of a zygote's combined genome. The final phenotype is formed only after the sequential realization of all the necessary genomic information.

Thus, the following conclusions can be drawn:

(a) all species of living organisms consist of individuals and each has a limited life term. Only the ability of individuals to propagate provides for the prolonged existence of a species (up to millions of years);

(b) all multicellular organisms consist of cells that vary in their structures and functions. The life of practically all cells is shorter than the life of the whole organism. Only a constant renewal of cells, maintained by their reproduction, provides for the prolonged existence of individuals;

(c) constant replacement of worn out cells with new ones provides a physiological regeneration of tissues and organs which conditions their prolonged functioning and maintains the integrity of the organism;

(d) the reproduction of organisms is a necessary prerequisite of evolution, since propagation is the basis of heredity and alteration;

(e) the process of reproduction conditions and maintains the phenomenon of generational alternation of both phenomes and their genomes, which is a mechanism of evolution;

(f) living bodies are not eternal, yet their genomes can propagate and travel through both time and space;

(g) periodic rewriting and editing of genetic information during each cycle of reproduction creates conditions for the strict control of integrity and absence of genomic damage;

(h) various living organisms have many different processes and mechanisms of reproduction. Nevertheless, the genome is the uniting factor for all of them.

Hence, it is evident that, in the process of reproduction, a trans-organismal movement of genetic information is carried out through special mechanisms of genomic transmission (Fig. 9.3). Thus, the main function of cells and multicellular organisms is to preserve and maintain structural and functional integrity for successful genomic transfer. Cells carry out the intra-organismal transfer through division, while the multicellular organism itself is responsible for inter-organismal transmission via reproductive processes. From this point of view, we may say that

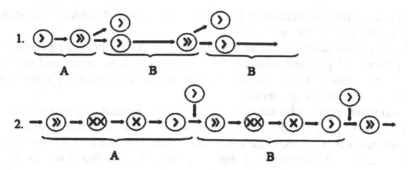

**Fig. 9.3** Reproduction is a transition of genomes through space and time via living bodies. *1*. A schematic representation of a trans-cellular transfer of genomes via binary division (unicellular), 2. A representation of trans-organismal transfer of genomes, including the processes of gametogenesis and fertilization (multicellular). *a*—maternal organism, *b*—daughter organism

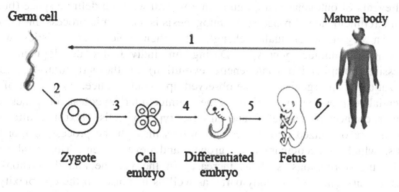

**Fig. 9.4** Meaning of individual development—creation of mature bodies that are producers, carriers, and disseminators of genomes. *1*—germ cell production, *2*—fertilization, *3*—division, *4*—differentiation, *5*—growth and development, *6*—maturation

reproduction is a necessary condition for the prolonged existence, not of various species of organisms, but rather of various species of genomes.

## 9.2 Individual Development

After fertilization, a wide spectrum of cells, tissues, and organs of a multicellular being is established from a single zygote through the mechanisms of cell division. The sum of consecutive processes from the moment of the appearance of a zygote to the termination of all living processes after death is called ontogenesis, or, more simply put, individual development (Fig. 9.4). The basis of ontogenesis is the selective realization of hereditary information at various developmental stages which feature discreteness and integrity.

The most important ontogenic characteristic of sexually reproducing organisms is the presence of germinal cells that form gametes such as ovules and spermatozoa. These cells and their derivative gametes are the means of continuous genomic transfer to new generations.

As mentioned previously, individual development is initiated in embryogenesis and propagated throughout the lifetime of an organism via the genetic programs of combined parental genomes. In the process of fertilization, the haploid genomes of a father and a mother unite into a diploid genome to be utilized by the future organism, although in the case of asexual reproduction (or parthenogenesis), the hereditary program is only contained in the genome of an ovule. During ontogenesis, the hereditary information stored in the diploid genome of an organism is gradually realized to yield specific morphological, physiological, and other features, which will become its phenotype.

All events of ontogenesis are tightly interwoven within a definite space (body) and time (coincidence of processes). Ontogenesis is a dynamic process throughout which an organism gradually changes its phenotypic characteristics while preserving its unique genotype. During an individual's development, the progressive complication (ontogenetic evolution) of the structural–functional organization of an organism can be observed up to pubescence. As a result of the differential expression of various genes, unicellular zygotes form nervous, muscular, connective, epithelial, and all other types of cells. Next, the differentiated cells are organized into tissues and organs through the process of morphogenesis, which is accompanied by growth and development. This vital stage includes transformations, such as increases in the volume and dimensions of tissues, organs, and various body parts, as well as increases in the complexity of their functions. Ontogenesis in multicellular organisms is a conditioned process whereby the state and completion of the preceding stages affect events that take place in subsequent developmental phases. For example, the limb formation begins only when an embryo achieves specific dimensions. Ontogenetic processes are clearly expressed in multicellular organisms, though some features are present in unicellular ones as well. In their life cycle, there are definite dynamic processes, such as those seen, for example, throughout the development of *Plasmodium vivax*.

Ontogenesis is a result of long processes of phylogenetic development. The reciprocal connection between individual and historical evolution is reflected in the law of biogenetics, which states that: "embryogenesis is a compressed repetition of phylogenesis." The process of ontogenesis is characterized by discreteness and integrity. Two distinct cells arise as a result of the first division cycle of a zygote, and these become the first components of a new biological system. From that pivotal moment, the development of an embryo is determined not only by genetic programs, but also by relationships between cells and elements of the materializing system. Every further stage of development (blastula, gastrula, neurula, etc.) is a new state of being for the entire organism. At any given stage of development, this system exists as more than a simple set of cells, because all of its cellular elements are deeply integrated and interconnected. This interaction and coordination of parts of the emerging embryo gradually changes during the developmental process, giving a sense of purpose to ontogenetic processes.

Ontogenesis can be divided into several periods:

*Pre-embryonic period.* This period is also called progenesis, and it precedes ontogenesis. The basis of progenesis is gametogenesis, which is the emergence and formation of mature gametes after the fertilization stage. The autonomous existence of gametes before fertilization is an intermediary stage of life and is the link between the ontogenesis of parents and the ontogenesis of their descendants.

*Embryonic period.* This period begins from the moment of an ovule's fertilization and amalgamation of parental genomes, which is the moment when a new body is generated. The period of embryonic development is divided into various stages such as those of the zygote, blastula, morula, gastrula, neurula, and the differentiated embryo. Before the emergence of organ precursors, the bodies of developing mammals are usually called embryos, and later, fetuses. These stages of development are conditioned by the cytogenetic processes of division, differentiation, migration of cells and cellular groups, histo- and organogenesis, growth, development, selective division, selective death of the cells of an embryo, embryonic induction, and so on. The end of an embryonic period is connected with the entrance of the organism into the environment through such processes as birth.

*Postembryonic period.* This period encompasses the premature and adult stages of development. The first stage is characterized by extensive growth, finalization of skeletal formation (if any), establishment of bodily proportions, and the completion of the development of a sexual system. The adult stage features the stable operation of an organism and an active production of gametes. Once they reach sexual maturity, pubescent individuals provide the necessary developmental conditions for the fertilization and unification of parental genomes. In other words, they fulfill their main intermediary role: they serve as a means for producing and disseminating genomes.

*Differentiation and morphogenesis.* Most multicellular organisms consist of many different types of specialized cells. Mammals, for example, consist of over 200 different kinds of differentiated cells. These cells have specific shapes and dimensions, localizations, and protein compositions, and they fulfill special functions. They are formed from non-differentiated embryonic cells through a gradual, genetically programmed change in their structure and functions during the process of development of an organism. Some cells remain virtually unchanged in the progression and some do not divide (e.g., leukocytes, macrophages, muscle cells, neurons, astrocytes, fibroblasts, and certain other cells). Differentiation is the process through which cells of a homogeneous germinal cell population acquire specialized and distinct morphological features, such as the formation of specialized structures. Similar types of these uniquely adapted cells join together to form specific tissues and organs. Differentiation is the result of a selective realization of genetic information through the "differential" expression of genes within a unified genome. The biochemical basis of the presence of various features is the synthesis and functioning of specific proteins which are necessary and unique to each type of cell.

Generally, embryonic differentiation of similar cells occurs in groups. Nearly simultaneously, all the grouped cells change their protein composition and the intensity of their metabolic activities. The composition of organelles, the construction of cellular membranes, and even the shape of the cells changes as well. These processes are accompanied by the morphological conversions of an embryo through morphogenesis, during which the precursors of organs appear and then develop gradually over time. The differentiation of cells and tissues limits the ability of different parts of an embryo to take other developmental paths. The process of obtaining neighboring cells of similar specializations via differentiation is called histogenesis. This allows for the formation of tissues that consist of analogous cells fulfilling similar roles and performing related functions. Thus, differentiation, histogenesis, and organogenesis take place simultaneously. However, due to selective gene expression, various regions of the embryo undergo completely different and independent processes of differentiation, and this demonstrates the presence of a strict coordination and integration of all the above-mentioned processes on the level of the whole developing organism. Finally, on the foundations of the differentiated embryo, a very complex and highly organized adult organism is formed after successive stages of growth and development, and this organism will in turn become a producer of gametes.

The expression of a gene in a trait is one of the main mechanisms of differentiation. Schematically, this process consists of several principal stages: structural genes → various messenger RNA specific proteins → cell differentiation → realization of specific functions. The extraction of genetic information is a very complex multistage process, which is controlled by a cell at several points, e.g., at the level of transcription, translation, and protein modification. These events establish a tight control over the differentiation process which continues well into adulthood, leading to the development of mature multicellular organisms. For example, hepatocytes (liver cells) express a specific group of genes that ensure the presence of proteins and enzymes specific for hepatocyte structure and functions and no other. These may include genes that code for such things as blood albumin and enzymes for glycogenesis, formation of urea, and so on. As another example, muscle cells express systems of genes for various contractile proteins such as actin, myosin, troponin, etc. Because of this, the specifics of the structure and functions of all organs and tissues are maintained during a prolonged period of time. Differentiation, therefore, is based on the sum of special genetically controlled molecular and cytologic processes which condition the directed development and structure of a complex multicellular organism.

*Growth and development.* All the stages of growth and development are under genetic control, and are maintained by a sequential and differential expression of genes throughout the process of ontogenesis. The central role of the processes of development in life is to provide translational mechanisms of virtual genomic programs in a specific physical organism. Through the genetically controlled use of matter and energy from the environment, consecutive events such as a rapid proliferation of cells, their differentiation, formation of tissues, organs, and the overall construction of the organism take place. All the stages of growth are coupled with a high metabolic activity of cells. An especially important role in growth belongs to the synthesis of proteins which comprise the main organic mass of cells. These proteins, be they enzymes or structural components, affect and drive cellular growth and development. The key programs of progression are rigidly fixed in the genomes of their hosts, and therefore persist through millions of years and generations.

Clearly, it is only by understanding the mechanisms of development that we can understand how biological structures and processes are formed, how they are operated by evolution, and finally, how they turn into the most complex organs, organisms, and species. The developmental pathways of living bodies from DNA to proteins have been rather well studied, and are known to involve such mechanisms as transcription, processing, translation, folding, and a number of other molecular processes which cause the formation of structurally and functionally active proteins. Despite this, most pathways encompassing such developmental events as the formation of macrostructures from proteins and the creation of entire autonomous living bodies are less well understood.

Thus, development is a totality of consecutive and irreversible processes of change in biological objects, which are normal and regular, and are particularly directed to cause the emergence of new qualities. The regularity of developmental events points to the fact that they are not merely casual occurrences that lie in the course of life, but rather that they are conditioned processes. The determinacy of these processes is connected with the material and informational essence of specific living bodies, particularly dealing with a genomic control of development which is driven by genetically targeted molecular and cellular interactions, as well as various environmental factors. The qualitative and quantitative changes associated with the processes of development therefore accumulate gradually and on a stage-by-stage basis. These changes manifest themselves through the emergence of a new system, as well as structural and functional properties of living organisms. From a thermodynamic perspective, the mechanics of the development of living systems involves the accumulation of negentropy and information. The set of information-saturated modifications gradually determines the evolutionary direction of biological objects and leads to a final result, which is the formation of phenotypes and the emergence of completed forms.

We can finally state that the process of individual development is a unique genomic mechanism involving the use of matter and energy to form living bodies. This mechanism is the basis for the emergence of new mature forms of multicellular organisms (Fig. 9.4). The latter produce gametes which are used to pass on the programs of development to new generations, thereby providing a genotypic continuity of life.

# Chapter 10
# Evolution

## 10.1 Adaptation

*Adaptation* is the phenomenon of organismal acclimation to the external world. As a result of the long concurrent development of the Earth and its life-forms, many very different habitats have appeared, each with its own unique set of inhabitants, underlying the general correspondence of specific living organisms with specific ecological conditions on the planet. These environments may differ in such features as temperature, aqueous states, atmosphere, gravitation, solar radiation, radioactive background, etc. The need of various organisms to survive and reproduce in their unique habitats has led them to develop specific adaptations such as special metabolic processes (for example, photosynthesis or oxygen breathing), functional abilities (ability to swim or fly), special organs (respiration, feeding, motion), and body parts (such as environmental sensors and various limbs).

Some types of adaptations occurred as a result of many consecutive mutations or recombinations of genes under the impact of environmental factors, which became fixed in the genetic apparatus of various organisms. Such adaptations generally develop over long periods of time, and bring significant changes in the genotypes and phenotypes of organisms. The emergence of new genetic alleles and their combinations conditions a change in the qualitative and quantitative composition of proteins, and this eventually provides a manifestation of new features, which may or may not be of any benefit. Generally speaking, those organisms that develop useful features continue to survive and reproduce, while the rest of the variants are annihilated as a result of natural selection.

*The process of adaptation*, therefore, is the change within an organism that tends to increase its chances of survival and reproduction. Since environmental conditions change over time, adaptations can also change, improve, or even disappear. Thus, adaptations are relative, because they appear in response to specific ecological problems. Under other ecological conditions, such adaptations may not fulfill any adaptive function, and with the disappearance of ecological

G. Zhegunov, *The Dual Nature of Life*, The Frontiers Collection,
DOI: 10.1007/978-3-642-30394-4_10, © Springer-Verlag Berlin Heidelberg 2012

pressures, the acquired features may turn out to be useless. Depending on the direction of changes, organismal adaptations can be differentiated by successive complications (for example, the appearance of a skeleton, ability to fly) or by simplifications of the structural and/or functional organization (for example, the loss of breathing organs and circulatory system by some internal parasites).

The environments in which living beings reside present not only material surroundings, but also a set of diversified spatial and temporal characteristics. Among them are permanent influences (gravitation and radiation), episodic influences (precipitations and earthquakes), and periodically repeated influences (seasons, sunrises, and sunsets). As a result of evolution and adaptation, all organisms reflect at least some aspect of these influences in their organization, in one way or another. For example, the force of gravity has caused the appearance of upper and lower body parts in many terrestrial and pelagic organisms. Furthermore, the periodicity of precipitations in some arid parts of the planet has caused the appearance of organisms that can periodically enter hypobiotic states during droughts, and then revive again during precipitations. The changing of seasons also provides a certain seasonality of reproductive and developmental cycles of many plants and animals. Prolonged phenotypic manifestations resulting from adaptive changes are caused by the selection of organisms that possess genotypes and phenotypes that are useful under given conditions. In other words, the phenotypes of all living organisms correspond to and reflect geophysical and environmental conditions, thereby ensuring comfortable surroundings for their unique genomes.

Adaptation is based on informational processes, because it is carried out during specific interactions of living bodies with environmental factors. Any changes in the material surroundings are perceived by receptors in organisms as specific informational signals. At first, functional systems that prevent or compensate for unfavorable factors switch on. However, if an irritating factor does not cease, those adaptations which are most adequate to deal with the given conditions gradually develop (over hundreds or thousands of generations) and are fixed in a genome. Therefore, the adaptation of the "phenotypic framework of a genome" is one of the main properties of the adaptation process which provides for specific inter-actions between organisms and their environment, and is the basis for survival and evolution. The qualitative and quantitative composition of all living bodies on the planet, their organization and evolution, and the origin of species can be considered a result of adaptations of the system of the Global Genome to the changing conditions of the environment throughout the development of the Solar System.

## 10.2  Heredity

The continuity of life on Earth is conditioned by the phenomenon of heredity. Heredity is a property of organisms to transfer and preserve similar features in a number of generations, and to provide specific characteristics for individual

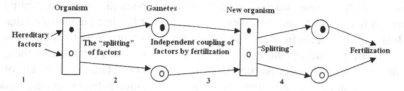

**Fig. 10.1** Schematic showing the inheritance of traits as suggested by Mendel. The main achievement of Mendel's work is the substantiation of the presence of discrete hereditary factors through which features are passed on from one generation to another. In other words, he was the first to hypothesize and prove the concept of a gene and its alleles as units of heredity. *1*—an organism contains two hereditary factors for any particular trait, *2*—during gametogenesis, they split and get into two different gametes, *3*—After fertilization, the factors couple again in the new organism, *4*—A new cycle of genetic transmission

development. It is due to heredity that parents and descendants share a similarity in the biochemical composition of their tissues, metabolism, functions, morphological features, and other traits. In other words, they inherit a genotype and phenotype which is standard for a given species.

Inheritance is a process for the transfer of similar features according to specific biological laws and across a number of generations. The transfer of such features and their properties is carried out by the process of reproduction. As mentioned previously, it is not the features themselves which are transferred, but rather genetic programs that condition their propagation. Genetic programs, in the form of various genes and genetic networks, are enclosed in a set of molecules of DNA, which are organized into chromosomes. Utilizing the gametes as mobile genetic carriers, the hereditary information can be passed on to later generations indefinitely by the process of reproduction. Therefore, every species of organisms reproduces practically unhindered for millennia without any major changes.

Many known patterns of hereditary features depend on the presence of various conditions. These conditions include such variable factors as which chromosome a gene resides in, whether they are allelic or non-allelic, how many genes determine the development of a feature, whether they are dominant or recessive, linked or not, etc. The first patterns of hereditary features were determined by G. Mendel around the middle of the 19th century and are summarized below (Fig. 10.1):

1. He defined the presence of hereditary factors in gametes through which features are passed on. (In the contemporary conception, a factor is one of the variants of a gene that determines a given feature—an allele). Every feature is controlled by two factors (two alleles). One of them can dominate (dominant allele) over another (recessive allele).

2. In the process of gamete formation, the "splitting" of factors takes place, and one of the two gets into each gamete. During fertilization, they can fuse into a zygote in different combinations, and therefore condition the development of different features.

3. Factors that determine various features are inherited independently from each other and can create many combinations.

Since Mendel's time, much new information has been discovered regarding the patterns of hereditary feature transfer. In particular, it has been established that the development of features in organisms is the result of complex interactions between genes and the products of their activity, namely, proteins. Furthermore, there are no features that arise under the impact of only one gene. The development of the majority of features stems from complex interactions between genes, genetic networks, and proteins, as well as the influence of various factors from the internal and external environments.

For example, the following patterns of heredity are well characterized:

1. Hereditary features are conditioned by gene alleles occurring in the same locations (loci) in homologous chromosomes.

   (a) *Complete dominance* manifests itself in those cases when one allele of a gene in a heterozygote completely conceals the presence of the second allele. This observation was quite characteristic for Mendel's experiments;
   (b) *Incomplete dominance*. There are genes that are neither dominant nor recessive. In such cases, both alleles control the manifestation of a specific feature. As a result, a hybrid that was received through interbreeding of two normal lineages of organisms does not resemble either of them, but is rather a mix of both;
   (c) *Codominance* is a kind of interaction of allelic genes in which heterozygous organisms show features that are determined by both alleles. For example, both proteins coded by these alleles are synthesized in such organisms.

2. The regularities of hereditary features are conditioned by non-allelic genes. Such genes are located on non-homologous chromosomes or in different loci of homologous chromosomes.

The overwhelming majority of organismal features result from the action of several different non-allelic genes. Such traits are referred to as *complex*. For example, a molecule of hemoglobin contains two $\alpha$- and two $\beta$-polypeptide chains. Their genes are located, respectively, on the 16th and 11th chromosomes. A complex feature is usually conditioned by the common action of several genes, or, more precisely, by a chain of biochemical transformations with the participation of many genes. Several examples of such interactions are presented below:

(a) *Complementary interaction*. In many cases, in order to form a feature, it is necessary to utilize an interaction of two dominant genes from various allelic pairs, because each complementary gene is not capable of providing the development of the given feature by itself;
(b) *Epistasis* is a type of gene interaction where one gene suppresses the action of another. Such suppressing genes are called inhibiting or epistatic. Genes whose expression is prevented by epistatic ones are called hypostatic. Epistasis

can be dominant if one dominant gene suppresses the action of another gene, or recessive, such as when a recessive gene suppresses the action of another gene.

(c) *Quantitative (polygenic) heredity.* Phenotypic features can manifest themselves in two ways: qualitatively or quantitatively. The development and manifestation of qualitative (monogenic) features are dependent on the expression of one allelic gene of a homologous pair. These features can refer to an organism's size, color, structure, sex, etc., and deal with the appearance of one of two distinct alternatives encoded by homologous genes at that locus. Quantitative (polygenic) features, on the other hand, do not exhibit distinct alternative forms, but instead a broad spectrum of phenotypes which can include many intermediary varieties compounded from alternative features. For example, quantitative attributes of a person's height, weight, skin color, size of organs, facial form, intellect, susceptibility to diseases, etc., are controlled by polygenic interactions and can be modified by environmental factors. Each gene brings an individual contribution to the phenotype, but only to some degree. The complex of two or more pairs of "plural" non-allelic genes creates an accumulating effect in the manifestation of a feature. The joint action of multiple genes conditions various levels of *expressivity*—varying degrees of the intensity of a feature—which depend on the quantity of expressed genes. The biological significance of polygenic heredity is apparent in the fact that there are a wide variety of features that increase organismal adaptation and contribute to evolution. Furthermore, features determined by multiple genes are more stable than those encoded by a single gene.

3. *Patterns of linked feature inheritance.* It is known that each chromosome contains many genes. During processes of cell division through mitosis or meiosis, all parental chromosomes are transferred as a whole unit, ensuring that all the necessary genes will be passed on as a coherent group. This property of chromosomes and their genes that they always stay together throughout processes of inheritance is known as gene linking. All genes situated on the same chromosome are called linked genes. The phenomenon of linking can be used to explain why, for example, a number of individuals of a population of *Drosophila melanogaster* (fruit fly) always have red eyes and gray bodies at the same time. The reason for the presence of this phenotype is that genes which determine these two features are on the same chromosome, which is packaged into a gamete during gametogenesis and then transferred to a new individual. Linking lowers the probability of the formation of new gene combinations in somatic cells and gametes, and therefore allows the preservation of parental, racial, and other specific features of the organism.

4. *Genetic imprinting* is a molecular genetic process that selectively turns off the expression of genes in the chromosome(s) of one of the contributing parents in a diploid organism. When this occurs, a mono-allelic (rather than a bi-allelic) expression of a gene can be observed, i.e., only the gene from a paternal or maternal allele is expressed. Imprinting constitutes a deviation from one of Mendel's laws which states that the contribution of each parent towards the

heredity of descendants is equivalent. Thus, phenotypic manifestations of a specific gene can change not only due to its mutation, but also as a result of the selective deactivation of its expression. The molecular mechanism of imprinting involves the specific methylation of cytosine bases, which subsequently turns off the transcription of the genes constituted by those bases. About 30 human genes are known to be susceptible to imprinting, and have a tissue-specific mono-allelic expression pattern. Some of these genes have a direct relationship to hereditary pathologies (such as tumors and certain syndromes). Genetic imprinting can be manifested not only at the level of a single gene or a cluster of genes, but can affect an entire chromosome, such as when a secondary X-chromosome in women is deactivated, or even a whole genome in extreme cases.

5. *Cytoplasmic heredity* is connected with the transfer of genetic information through DNA-containing organelles such as mitochondria and chloroplasts (plants). These have their own DNA which can also be inherited during cell division or reproduction.

It is important to emphasize that, despite multiple varying factors such as complex inheritance mechanisms, gene interactions, cross-over events, and the intricate processes responsible for the formation of features during development, it is nevertheless possible to form basically identical individuals which possess the same morphological, physiological, biochemical, and genetic characteristics as their predecessors. This inherently unwavering stability conditions the genetic continuity of individual species by allowing their genetic information to be repeatedly transcribed, passed on, saved, and reproduced, practically without any major changes in their genomes or phenomes for hundreds of millions of years.

Thus, heredity is a very important phenomenon which is connected to the diversified and complex interactions between discrete elements of an organism's genetic apparatus. Proper hereditary processes depend upon the necessary presence of several key factors, such as the high stability of genes, DNA molecules, chromosomes, and karyotypes of individuals, as well as the high accuracy of molecular and cellular mechanisms that provide a transformation of genetic material and information through the process of reproduction.

In recent years, a new mode of transfer of hereditary information has been established, one which apparently does not require the participation of nucleic acid molecules. This phenomenon was discovered in connection with several diseases including "mad cow" disease, Creutzfeldt-Jakob disease in human beings, Scrapie in sheep, etc. It was shown that these diseases occurred due to protein agents called prions which affect the nervous system. It was established that prion proteins are absolutely identical in their amino acid composition to normal endogenous proteins, but adopt strikingly different conformations which then persuade other normal proteins of the same type to become abnormal. Such anomalous proteins tend to coagulate and form aggregates (amyloid), which interfere with normal neural cell functions. Thus, prions represent the transfer of information from one type of protein to another without the participation of nucleic acids! It is as if the anomalous proteins "transfer" their conformation directly to normal proteins. This type of information transferring mechanism (concerning protein spatial forms) is called protein heredity.

It is important to note another method of storage and transmission of non-genetic hereditary information. This method deals with the tremendous mass of information which is present in the form of the highly organized molecules and supramolecular complexes of the cytoplasm. Cytoplasmic order is a prerequisite for the operation of the colloidal matrix found within cells, which conditions identical instruments and mechanisms for the transformation of matter, energy, and information. The orderliness of molecular systems and supramolecular complexes is always evenly distributed in daughter cells when dividing (cloning), since the transfer of genetic material is not enough to initiate its processing in the newly established cell. This cytoplasmic transfer is merely for convenience, because there is then no need to build this order anew in the new cell. When considering heredity from this perspective, the DNA molecules embedded in the cytoplasmic matrix can be considered only as an equivalent part of the overall system of life. Therefore, it is important to remember that heredity is not merely genetic, but also occurs through structural heredity, or the expedient transfer of cellular order.

It is also important to take into account mechanisms of epigenetic heredity, which are connected to the inheritance of phenotypes arising from the selective modification of genetic states without changes in the basal DNA sequence. One of the best-studied epigenetic features which occurs during cell division is the methylation of DNA and the modification of histones, which establish a definite structure of chromatin and allow for the transformation of genetic information encoded there. Furthermore, it is well established that a DNA molecule is irregular, and its local conformation is dependent on the nucleotide sequence, which leads to the observed structural variations such as those found in the small and large grooves, the twisting of the bases of neighboring pairs, and other parameters. This is the basis for the "conformational DNA code", which determines specific localization sites recognized by regulatory proteins.

With the help of such mechanisms as the ones just mentioned, features are inherited which are not connected with changes in the genetic code. It therefore becomes evident that there is no monopoly of DNA in the transfer of hereditary information. All the other cellular components, in collaboration with DNA, are also responsible for ontogenesis, differentiation, structures, and functions inherent in cells.

All cellular processes are regulated and carried out through the synthesis of special proteins. Besides genetic regulation, this can be controlled and managed at the cellular level through such mechanisms as alternative splicing, post-translational modification, folding, etc., conditioning the ability of a single gene to serve as the basis for the formation of different proteins.

It should be clear from this discussion that the transfer of hereditary information can be naturally divided into genetic means (connected with NA) and non-genetic means (unrelated to NA). With respect to the genotypic and phenotypic levels of life, the various mechanisms of inheritance of genetic material can be defined as "genotypic inheritance", while the diverse mechanisms of inheritance of non-genetic information can be categorized as "phenotypic inheritance."

Nevertheless, regardless of the nature of inheritance, it is a general phenomenon which is "inherent" to all living bodies and is exceedingly important. Inheritance permits the transfer of organizational and functional blueprints to millions of generations of organisms, providing for the long-term existence of diversified species of living beings as well as the continuity of the phenomenon of life on Earth.

## 10.3  Variability

Variability is the property of living organisms to obtain new features or new types of features (which can arise in one or more new combinations), or to completely lose certain features. Variability generates an enormous amount of features and their variants, and this conditions processes of adaptability and survival in organisms under constantly changing environmental conditions, as well as generating a tremendous variety of organisms. Because of the variability within any population, every individual possesses only its own unique combination of traits. It is because of this that people, for example, have different colors of eyes, hair, skin, nose shapes, ears, body dimensions, blood groups, temperament, intellect, susceptibility to various diseases, etc.

If the new features are transferred to offspring, then this type of variability is termed hereditary or genotypic variability, but if they are not passed on, it is non-hereditary or phenotypic. We will now discuss some different features and mechanisms of variability and its consequences for development and behavior.

1. *Phenotypic (modification) variability* is the ability of an organism to change its phenotype under the impact of factors from the external and/or internal environment without the transfer of these features to progeny. Phenotypic alterations that appear under the influence of environmental conditions are called modifications, and are generally only concerned with the features of somatic cells. The intensity of modifications is proportional to the force and duration of the action of a given factor. If the conditions of the environment act upon the organism during critical periods of development, they can considerably change the normal flow of developmental events, and this can lead to abnormalities or death. New features appear due to the fact that physical and chemical factors of the environment act upon physiological and biochemical processes that take place in an organism, and thereby change their flow. In most cases, the modified variability results in an adaptive reaction of the organisms which is directed to ensure their survival under specific conditions.

In many cases of environmentally induced modification, there can be changes in the genotype of somatic cells. These are called *somatic mutations*. Because these mutations do not occur in germ line cells, they do not result in genotypic changes in descending generations of individuals, since they are not inherited. However, for the cellular offspring of such mutated cells, the variability is hereditary, because later generations of cells will inherit and then clone this mutation. That is how cancerous tumors can develop in a body, because they are clones of the same mutated somatic cell, which gain properties that tend to allow them to proliferate indefinitely. In other words, the same mutation can be hereditary in one case (for a line of cells) but not in other cases (for organisms). It is therefore clear that there exists a conditional character to the division of variability into hereditary and non-hereditary forms.

Mutations in somatic cells may be present in several forms and locations: genes, chromosomes, and whole genomes. Such mutations may be caused by physical, chemical, or biological factors, and may be the reason for the appearance of new types of cells, tissues, or organs. Such genotypic modifications are only heritable to cellular progeny within a single organism. Because these mutations are heritable to future cells, they may result in detrimental consequences. For example, as a result of changes within DNA, a group of cells can arise which may have defects in the regulation of cellular reproduction and thereby gain the ability to undergo unlimited division.

The limits of modification variability are confined within genetically conditional norms of reaction. The norm of reaction is a range which defines the prevalence of feature modifications in response to the changing conditions of the external or internal environment. The reaction norm is adaptive and forms gradually as a result of natural selection, developing in compliance with different living conditions. Natural selection favors genotypes with a broad norm of reaction, since such organisms have better chances of adequately responding to the influences of various environmental factors.

Practically all features have varying degrees of manifestation and expression in the individual organisms within each species, a characteristic of feature polymorphism. It is the presence of different alleles in the cells of individuals, differential gene interactions, and the influence of internal and external factors that provides for the tremendous variety of combinations of features and their variants, e.g., the multitude of colors and shades of human skin.

*Ontogenetic variability* is a type of phenotypic variability. This type of variability is connected with specific stages of individual development in the process of ontogenesis, and covers considerable phenotypic changes without altering the genotype. Due to the differential expression of genes at all phases of development, only the necessary genotypic segments function at each stage. This conditions the definite differentiation and morphogenesis of cells and explains why organisms have certain sizes and body shapes, as well as the structures of inner organs throughout development. Animals that develop through metamorphosis, such as the butterfly, demonstrate completely different phenotypes at immature stages (i.e., pupa) compared to the adult organisms, even though both are genotypically identical. Even human beings look different at various developmental stages. This is most obvious during early stages of embryonic development, when a single cell

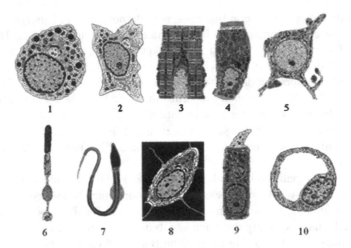

**Fig. 10.2** Interclonal variability of human cells. The basis of interclonal variability of cells depends on the differential expression of genes in various cells of tissues and organs, which are all united by the possession of identical genomes. Several examples of the many different cell types which arise from a single zygote cell are presented here. *1*—myelocyte, *2*—fibroblast, *3*—myocyte, *4*—epithelial cell, *5*—neuron, *6*—rod cell, *7*—sperm cell, *8*—osteocyte, *9*—odontoblast, *10*—endothelial cell

transforms into an embryo that has hundreds of diversified cells and all kinds of tissues and organs. The order and direction of ontogenetic modification is exceedingly stable and unidirectional, and may not change for millennia, since the programs of development are determined by the genome.

*Interclonal* variability can be distinguished as a specific type of phenotypic variability which accounts for the presence of various cells within a single organism (Fig. 10.2). All cells of an organism originate by clonal division of a single progenitor cell (zygote), and have the same genome. Most adult animals, especially mammals, consist of over 200 phenotypically different types of cells that fulfill numerous different roles. The basis of such interclonal changeability depends upon the differential expression of genes in different tissues or organs. Cells from different lineages tend to express specific proteins that can create the peculiar conditions determining cellular fates. For example, the cells of muscle tissues express large amounts of contractile proteins such as myosin, actin, troponin, etc., which may never form in other cells.

2. *Genotypic (hereditary) variability* is connected with changes in the hereditary apparatus of germ line cells, which can be transferred to later generations of organisms. This type of changeability is highly significant because it serves as the primary source of genotypic and phenotypic diversity in offspring, and this in turn leads to the processes of natural selection and evolution.

*Combinatorial variability.* This type of variability appeared alongside sexual reproduction and is caused by recombination events both between and within the alleles of parental genes in germinal cells. It serves as a source of endless combinatorial variants of different features. That is the reason why there are no organisms with absolutely identical features (except for monozygotic twins, though even they differ in their sets of somatic mutations). Combinatorial variability is conditioned by the following processes: the countless variation in (a) the meeting of sexual partners who have different genotypes and phenotypes; (b) fertilization of unique ovules with spermatozoa; (c) allelic recombinations during crossing-over; and (d) independent and random distribution of chromosomes in meiosis.Because combinatorial variability allows for natural selection, it plays an important role in the appearance of new populations, a feature often exploited by humans. Many different types of cultivated plants and breeds of domestic animals were created by hybridization of breeds that existed at earlier times. Combinatorial variability, therefore, is a very significant factor in the processes of adaptation, evolution, and speciation.

*Mutational variability.* Mutation is a sudden stable structural change of genomic elements which causes the appearance of new features or new variants of features in cells and is characteristic for all gene-based organisms, including viruses. Mutations can be useful, neutral, or harmful. If harmful mutations appear in generative cells, they could be transferred to offspring, and while they are generally not dangerous to the gamete producers, they may be hazardous to the posterity of their offspring. Hereditary anomalies, such as problems in the construction of the body, failures of various functions, and diseases, can appear in the progeny as a result of such mutations.

The process which results in the appearance of mutations is called *mutagenesis*, and the factors that cause mutations are called mutagens. Two mutagenic factors can be distinguished: exomutagens—factors of the external environment, and endomutagens—factors of the internal environment, which are metabolic products such as the toxic forms of oxygen produced by the action of the electron transport chain. Exomutagens can be (1) physical (ionizing radiation, ultraviolet rays, temperature, etc.); (2) chemical (formalin, mustard gas, colchicine, resins, heavy metals, medical substances, toxins of bacteria and parasites, etc.); and (3) biological (viruses, viroids, plasmids, episomes, and IS-elements). Mutations can also arise due to mistakes in the replication and/or recombination of DNA, as well as through problems in mitosis and meiosis. The cells (or organisms) in which mutations occur are called mutants.

Mutations and genetic recombinations are occasional and unpredictable, as is the entire process of heritable variability. A great number of different physical, chemical, and biological factors can affect entire chromosomes, specific chromosomal segments, or even individual nucleotides within germinal and somatic cells. Recombinations can also occur in any of the above-mentioned regions of the genome. The compounding of such events leads to the formation of diversified phenotypes within a given species, only some of which end up being chosen to persist through the sieve of natural selection. As a result, there have been and still are many organisms that have gone through the natural selection process and have adapted to specific conditions over many hundreds of millions of years of evolution. Nevertheless, there are much more frequent and random mutations that occur in somatic cells and are easily eliminated merely by the death of the organism.

## 10.4 Evolution

An endless multitude of living organisms has developed from simple unicellular primogenitors through the process of evolution. *Evolution* is a continual process of progressive changes of living bodies according to specific laws of development under the impact of environmental forces. Although evolution continues all the time, it is also quite gradual, and the time frame in which organisms become increasingly adapted and complex may be on the order of millennia.

In many cases, lower organizational states are preferable to higher ones (for example, in internal parasites). Adaptation, whether it involves a gain or a loss of function, is a directive force in the development of life, as are directed genomic changes. The appearance of any specific mechanisms of adaptation depends on living conditions, which may elicit several different adaptive mechanisms in organisms. For example, in order to survive in low temperature conditions, organisms have developed a multitude of complex mechanisms including anabiosis, winter hibernation, thermal regulation, migration, antifreezing, etc.

According to Darwin and Wallace, natural selection is one of the most important mechanisms of evolution and the emergence of new types of organisms. Natural selection allows for the death of individuals that failed to adapt, while promoting the survival and reproduction of those that did. Those organisms that have survived possess better features with corresponding genotypes, which are transferred to future generations. A newly created population, and later a whole species of organisms, can therefore successfully exist in newly altered conditions. Darwin's theories rather convincingly explain the variety and source, as well as the mechanisms of appearance and development, of various living bodies. It can also be added that Darwin's postulates were based, as we now understand, on the principle of the continuity of the processes of life's development.

The modern understanding of evolution is called the synthetic theory of evolution (STE), which is a theory of the evolution of organisms through a natural selection of useful features that are genetically determined. As mentioned previously, under the impact of environmental factors, DNA mutations and structural changes occur in some genes, and this causes altered proteins to be synthesized in cells that stipulate the development of new organismal features. DNA recombination and gene exchange also make significant contributions towards the appearance of new features. If these features raise the chances of survivability and adaptability, then they are more likely to be passed on to later generations. Thus, naturally, useful genes are selected through the mechanism of phenotypic selection, causing new genetic networks to be built, which in turn consolidate the adaptation and result in evolution. This general evolutionary theory combines Darwin's stipulations with the positions of modern genetics and molecular biology.

Cytogenetic mechanisms establish the foundation for all continuous global processes of evolution and development, and include systematic processes such as: (1) mutation; (2) combinatorial variability; (3) transgenesis; (4) hybridization; (5) symbiosis; and (6) natural selection of genetic information.

The consequences of changes in the genome are intensified many-fold in the processes of protein synthesis and the division and differentiation of cells. This in turn results in the appearance of conditioned features and their diversified combinations, i.e., new variants of phenotypes. At this stage, natural selection of favorable phenotypes takes place. Such organisms survive and contribute to posterity, securing useful features in subsequent populations. Therefore, the changes in the genetic apparatus are intensified in the process of phenotypic formation, and then undergo natural selection by environmental factors.

All of these processes can really be summarized by three distinct fundamental biological phenomena: variability, heredity, and natural selection. Variability, based on the plasticity of genetic material, creates a number of variants and pathways of development in a given biological system; heredity, which is conditioned by the conservancy of a genotype, narrows down the number of variants, whereas selection, in compliance with environmental conditions, secures a limited number of adapted forms for propagation.

An important postulate of STE states that an evolutionary unit is a population, where the frequency of specific gene changes determines the prevalence of a new feature. Individual organisms do not tend to evolve because they are already developed and have definite lifespans. Nevertheless, over the course of many generational transitions, a population may accumulate a given set of mutations that may establish new gene combinations and subsequent changes in the genome. As a result, more individuals would appear that possess new functional properties. The result of the evolution of a specific population is, therefore, the appearance of a new species, or, rather, of a new general genome. In our view, this means that a unit of evolution is the cumulative genome of a population. It is the very part that experiences primary modifications, subsequent to which its phenotypic framework changes.

Evolution is characteristic for all organizational levels of life and consists of several universal tenets: (1) evolution of organic molecules has led to the appearance of vital macromolecules; (2) evolution arising from various unions and interactions of macromolecules caused the formation of cells and cellular organelles; (3) evolution of cells led to the appearance of thousands of structurally and functionally different types of cells, tissues, and organs; (4) evolution of different types of cells and tissues led to the growth of several fundamentally different kingdoms of living organisms which have been covered previously; (5) within every kingdom, depending on the surroundings, individuals developed in various directions, forming a multitude of species that differ in their morphology and physiology.

The amazingly appropriate organization, interaction, and interdependence between various organisms can be explained by the process of *co-evolution*. Through this process, living bodies simultaneously changed and developed for billions of years by constant interactions, gradually creating, selecting, and layering specific convergent and divergent processes, structures, and mechanisms. For example, molecular co-evolution has given rise to the structures of DNA, RNA, the language of the genetic code, processes and mechanisms of transcription and translation, and the unique set of molecular enzymes which fulfill unique and highly specialized roles. The processes of parallel and reciprocal co-evolutionary accommodation have led to the ideal interactions of various cells in multicellular organisms, and to the co-development of multitudes of organisms in specific ecological niches throughout millions of years and generations. For example, co-evolution can explain the existence of strict feeding chains, the breaking or failure of which can lead to the deaths of scores of individuals from various cohabitating species. Other examples of this phenomenon are clearly observed in the reciprocal dependence between insects and pollinating flowers, as well as between certain parasites and their specific hosts. Plants, animals, bacteria, and other organisms have evolved simultaneously and interdependently as the derivatives of the Integrated System of the Global Genome. It is therefore apparent that no population within a specific ecosystem evolves independently, but only through its interactions with other populations and species, as well as with the environment. We can say that co-evolution is a global phenomenon of living Nature that underscores integral and interrelated processes of modification of all genomes (GG) and phenomes (GP) of the Integrated Life System, whereas evolution corresponds more to the processes of alteration within a concrete species.

The geochemical and geophysical conditions on Earth have been altered many times and by significant degrees within the period of its existence. From the moment of life's appearance to the present day, millions of species of living organisms have appeared, existed, evolved, and died on this planet, and their remains are constantly being discovered by paleontologists. Any change in physical conditions, no matter how big or small, causes modifications in the development of living organisms, and these in turn subsequently change the Earth and its conditions through reaction, resulting in a constant reciprocal loop.

Thus, a number of consecutively large evolutionary changes bear evidence to the progress of life's global development. Having appeared as simple life-forms, organisms developed with an increasing complexity, establishing refined functions and high levels of adaptability. Every new species of living organisms that appears on Earth is not maintained for eternity, but only for as long as there are environmental conditions that can satisfy its specific life requirements. At any stage, therefore, evolution has an adaptive and temporary character.

If we consider evolution in retrospect, it is evident that, as we descend the evolutionary ladder, we get ever closer to the fundamental essence of life (Fig. 10.3): from multicellular to unicellular organisms, from eukaryotes to prokaryotes, and from prokaryotes to viruses. As a result, the "naked truth of life" is that it is basically a minimal complex of nucleic acids (DNA or RNA) and several proteins. Interacting combinations of these molecules determined the appearance of qualitatively new systems which led to the development of living bodies. It is interesting to note that the critical molecular processes that condition

**Fig. 10.3** Evolutionary stairs that lead from ancient times to the unknown future. One of the greatest intellectual achievements of human beings is the understanding that an endless multitude of species of living organisms developed from simple unicellular primogenitors, that every new species is strictly adapted to its surroundings and the way of life, and that the process of evolution is continuous and directed towards the future

life have not really evolved to any great extent. The key processes of replication, transcription, translation, and enzymes that serve these mechanisms, are practically identical between modern multicellular organisms, including humans, and prokaryotes, which have lived for over 3 billion years.

Evolutionary processes in biological systems are probabilistic in character, and have been for billions of years. As yet, there is no evidence of the existence of some specific plan in the global process of evolution. DNA can have a dramatic influence on the order of the surrounding chaotic material space. Nevertheless, DNA itself also appeared as a random happenstance of chemical evolution. It is therefore subject to the same probabilistic influences which result in mutations, which are in turn selected by environmental conditions. In other words, the appearance of living organisms and their various features is determined by random actions of specific factors of the environment on nucleic acids. Because evolution is constant, random, and repetitive, organisms will continue to develop in absolutely new and different ways. In the case of a possible repetition of evolution of some specific living organisms, the process could flow in a totally different and unpredictable direction. Therefore, each type of living organism is unique and, once it has disappeared off the face of the Earth, it will never reappear again.

It is considered that the various species of living organisms gradually evolve over rather long periods of time, ranging from 2–10 million years, coinciding with changes in the environment. After a round of evolution, a species may either continue to be modified or disappear altogether. If environmental changes are dramatic and relatively quick on the evolutionary scale (occurring within several thousand years or less) the species will very likely disappear due to an inability to adapt this quickly. It is possible that trilobites, ancient fish, dinosaurs, giant birds, mammoths, and other large species, classes, and types of living organisms disappeared suddenly in this manner. In other words, we can say that evolution is not just the process of gradual adaptation and modification of species, but also quick saltatory destructions of old genomes. Probably, both gradual and saltatory mechanisms of extinction and emergence of new species are typical for the global process of co-evolution.

The synthetic theory of evolution considers the genome as a passive structure that encodes and transfers randomly appearing genotypic variances, and constantly sifts through them by the process of natural selection. According to this point of view, only selection actively transforms random genotypic modifications into necessary features, while the genome merely follows the selection in a passive manner. Nowadays, however, more and more experimental facts point to the possible existence of independent and purposeful genomic modifications that appear not to be random in nature. These new facts attest to the idea that the genome is an active, driven, self-regulating, and self-organizing system, which includes not only the materials necessary for evolution, but also the mechanisms of its use in specific and non-random ways. Even at the end of the last century it was assumed by Barbara McClintok, who discovered transposons, that the "genome is a highly sensitive organ of a cell, which, under stress, can initiate its own re-organization and updating". It appears that there are mechanisms in the genome that create and control purposeful, diversified, and coordinated alterations of the DNA. It is quite possible, therefore, that evolution is stipulated not only by changes in the conditions of existence and adaptation, but also by relatively independent processes that are based on internal laws of genotypic life, the mechanics of which is not yet clear to us.

Let us take a look at the process of evolution from the perspective of our conception of the coexistence of genotypic and phenotypic life, which implies that the entire aggregation of living organisms on Earth is considered to be the Global Phenome, and is a direct derivative of the expression of the Global Genome. If this is so, then the selection of phenomes is a secondary mechanism of evolution. The primary causes of evolution, which condition changes in the system of phenotypes, appear to be the molecular alterations of discrete genomes that compose the Global Genome, these being provided by the processes of hybridization, mutation, and recombination, as well as through the activity of mobile genetic elements. The changes that appear in the NAs can then be modified, extracted, and transferred into any other discrete genome within the Global Genome with the help of the same mobile genetic elements. Viruses and their derivatives (plasmids and episomes), along with molecular mechanisms of NA manipulations, are actively participating instruments in the endless evolutionary procession. Thus, the information-material system of the Global Genome exists and reacts as a single unit. It contains all the necessary molecular mechanisms and instruments for the perception, processing, and transfer of incoming information as molecular and atomic fluctuations of NAs. It is these processes that are primal in the informational system of the GG. It is in the system of the GG that there lies a still obscure parallel world. Expression of modified discrete genomes leads to the appearance of new phenotypes which go through selection. Thus, it is possible to evaluate the evolution of the GG by modifications of its expression during billions of years into new, more complicated phenotypic forms. Random as well as directed modifications of concrete genomes represent molecular mechanisms of evolution.

Therefore, the obviously interrelated modifications in the totality of phenotypes of living bodies are ultimately just a consequence of the co-evolution of discrete genomes within the integral GG system.

The global molecular evolutionary network is supplemented by other, more abstract forms, namely scientific, technical, and cultural evolution, which concern only human beings. The main aspects of this type of evolution deal with the application of science and the utilization of scientific data in everyday life, such as through the spread of computer technologies and the manipulation of genetic materials. Likewise, the process of social evolution is connected with the transfer of cultural and scientific information from one generation to another by non-genetic means. This type of social information evolves as well, and also passes through natural and artificial selection based on the criteria of its usefulness to individuals and society. Social information is purposefully created by human beings, whence a specific rationale underlies its existence. In other words, taking into account the significant influence of human beings on the world, we can say that humanity has become a powerful new factor in the evolution of both animate and inanimate Nature.

New species of living organisms appear, live, progress, and finally either modify into forms better adapted to new environments or vanish as a result of an inability to survive in new conditions or an inability to compete with stronger species. No organisms, including man, are guaranteed eternal prevalence on this planet. It is entirely possible that in thousands or millions of years, humans will be unpredictably modified or even disappear completely, because it is difficult to imagine the types of changes that may occur within our genomes due to environmental conditions or new competing species. It seems that while the nature of evolutionary causality is quite unknown, the same can be said about the reasons for the appearance of matter and the Universe itself. In the words of Charles Darwin, to "ask about the origin of life is simply stupid, like asking about the origin of matter." This statement implies that the main unresolved problems of both the past and the present are rooted in the conundrum of the presence of an initial unknown impetus.

According to the second law of thermodynamics, the Universe, and by association everything in it, tends towards greater chaos and randomness. Therefore, the strategy of the development of life on Earth has been driven by the ability to evolve in order to survive. Thus, entropy is the ultimate reason for evolution. Evolution is inevitable and continuous due to the fact that newly appearing organismal adaptations are imperfect, and sooner or later are surmounted by the pressures of entropy. If such an organism wants to survive, it must continue to evolve. The choice is evolution or death!

Despite all Nature's beauty and majesty, it must be remembered that it is merely the result of thousands of manifestations of stochastic events caused by the boundlessly evolving material world during the time span of the Earth's existence. Thus, all the wonders of Nature that we know today are merely temporary facades of ever-changing internal processes confined in a given space and time, which includes everything that is alive. We can therefore postulate that the only thing which remains unchangeable and unwavering is change itself.

# Chapter 11
# Homeostasis and the Maintenance of Integrity

## 11.1 Homeostasis

Organisms exist as complex, relatively stable physical and chemical systems, despite the permanent wear and degradation of particular elements. Such stability is manifested through various factors including the permanency of their molecular composition, acidity, temperature, pressure, and so on. The ability of an organism to maintain its integrity secures its survival and prolonged existence under variable conditions. All organisms developed anatomic, physiological, biochemical, and behavioral adaptations that serve one purpose: to preserve the constancy of the internal environment in order to provide optimal conditions for vital activities and reproduction. Organisms are able to maintain and regulate all structural and functional parameters (within certain narrow limits) for quite long periods of time, a necessary condition for uninhibited life.

The self-maintenance of the dynamic permanency of an organism's internal environment through the stability of its chemical composition, physical and biological properties, metabolic stability, and physiological characteristics is called *homeostasis*. Homeostasis provides organisms with a relative independence from an external environment, as well as a permanent level of activity despite variations in the conditions of the external surroundings.

To ensure active stability, homeostasis is generally maintained at all organizational levels, which for complex multicellular organisms such as animals, includes molecules, cells, tissues, organs, systems, organisms, and even populations. Human beings, for example, have hundreds of controlled homeostatic indicators. At the molecular level, a constant number of chromosomes, the consistency of genes, the stability of the genetic code, and the composition of nucleic acids secure genetic homeostasis. At the next level, cells control their structural and functional order by regulating the qualitative and quantitative composition of enzymes, an optimal set of ribosomes, mitochondria, etc. Higher still, at the organismal level, biochemical and cytological conditions of the blood, the pH of cells and fluids, hormone levels, as well

G. Zhegunov, *The Dual Nature of Life*, The Frontiers Collection,
DOI: 10.1007/978-3-642-30394-4_11, © Springer-Verlag Berlin Heidelberg 2012

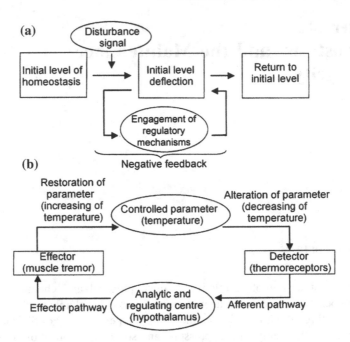

**Fig. 11.1** Principles of control for homeostatic characteristics of the phenome. **a** scheme of regulation of homeostatic characteristics. **b** components of negative feedback in the case of thermoregulation

as respiration, temperature, and many other parameters are constantly maintained and modified in response to changing situations, thus ensuring the continuation of life.

The aim of homeostasis is not to maintain a rigid set of parameters, but rather a working range within which minor fluctuations are allowed in the systemic balance, thereby permitting the organism to respond to changes. Under physical pressures or some external or internal environmental factors, homeostatic parameters can temporarily vary, but only within the permitted range. When such pressures cause the organism-environment relationship to become unbalanced, adaptational mechanisms first compensate for the changes in the physiological state of the being, and then attempt to return the system to a state of dynamic balance (Fig. 11.1). Regulatory mechanisms which manage homeostasis are based on principles of automatic control and correction of counteracting and restorative processes, a characteristic feature of closed-loop cybernetic systems.

The ability to maintain the constancy of the internal environment depends on the level of evolutionary development and the peculiarities of phenotypic elements that have direct contacts with the internal and external environments. In particular, this includes the ability of specific structures to limit fluctuations and return metabolic parameters and functions back to normal levels. In mammals, processes involve the coordinated efforts of many cell, tissue, and organ structures. For example, in order to return an elevated concentration of glucose in blood to normal levels, it is necessary to have a working detector that can "sense" the amount of glucose in the blood, a task that is achieved by special protein receptors on Langergance islets of the pancreas. It is then important to transmit this information to $\beta$-cells, which activate the process of insulin release from secretory vesicles into the blood stream in order to trigger glucose uptake by various other cell types. Concurrently, insulin genes are activated, causing an accelerated synthesis of new hormone molecules. This leads to the formation of new vesicles with insulin, and these are subsequently transported to the plasma membrane and released into the blood stream. Thus, in order to just maintain glucose homeostasis, an organism engages many diverse systems which act together to achieve the desired task. Such systems include the pancreas, the $\beta$-cells of Langergance islets, membranes, receptors, genes, and numerous other enzymes. All these regulatory units are interrelated elements of a single phenotype, the unique peculiarities of which depend on the genome. Thus, it is obvious that homeostatic characteristics and the ability of organisms to maintain them were formed in the long process of evolution, are genetically conditioned, and are directed toward sustaining genomic life.

Homeostasis is a necessary requirement for the existence of any organism, from unicellular life forms to human beings. It is a specific and directed process which is maintained by the permanent influx of matter and energy into the biological system through metabolic processes, and it is under the direct control of the genome. In conclusion, we can say that organismal and cellular homeostasis maintains comfortable surroundings for the genome which provides for all vital processes.

## 11.2  Confined Existence

The wonder of life carries with it several harsh realities. Nothing is eternal. Everything is destroyed according to the laws of physics. Everything material has a limited existence. Entropy is relentless and it concerns living bodies just as it does everything else.

*The duration of cellular lives.* Complex organisms, such as mammals, consist of an enormous number of cells forming different tissues and organs. The majority of cells live only for a specific period of time because they get worn out, grow old, and eventually die. Their place is taken by other cells that arise due to division. In other words, organisms possess an ability to restore and maintain their cellular composition, and this provides a certain degree of integrity and specificity to the duration of individual lives. Because many cells have different lifespans while having the same genome, it is evident that the overall interval of life is determined genetically. For example, many cells have a genetic program that triggers their timely self-destruction (apoptosis), and this is activated in the case of infection, wear, injury, or genetic degeneration.

Cellular macromolecules, such as DNA, RNA, and proteins, also have well-defined lifespans. DNA molecules are fairly stable and are therefore long-living due to several key repair systems. RNA molecules generally have short lifespans because they serve as mediators in the protein synthesis cascade, and are therefore quickly regenerated. Proteins are rapidly worn-out due to deleterious effects from such factors as thermal motion, chemical alteration, toxicity, and radiation, and therefore only exist long enough to perform their specific roles. An average lifespan for cellular proteins would be several hours. Nevertheless, on the basis of genetic programs, new proteins are constantly being re-synthesized, allowing the cells to live considerably longer than their proteins. The absence of some proteins, errors in their synthesis, and unexpectedly harsh wear may result in abnormal cellular functions that can lead to premature cell death. Thus, the stability of the qualitative and quantitative protein composition of a cell has a profound effect on its lifespan.

The lifespan of cells is not connected with the lifespan of their genomes. In the case of successful division, the parent cell may be slated to die, but the duplicated genome continues to live on in the phenotypic framework of the daughter cell, until it too suffers the same fate. And so it continues until there arises a given species.

*The duration of organismal life.* Millions of different types of living beings exist on the planet, each of which has a genetically programmed maximum term of life that is generally only reached by a few individuals. For example, for most rats, the lifetime is 5 years, for dogs it is 25 years, for chimpanzees it may be 60 years, for elephants it is 70 years, and some human beings can live for as long as 120 years, though the majority do not. Many individuals do not live as long as their programming would otherwise allow because of physical problems such as heart murmurs, ailments such as cancer, and accidental death from any number of factors.

However, it is still unclear what mechanism determines the term of life. Why do some species, despite being lower on the ladder of complexity and evolution, appear to live much longer than more advanced and developed species? Many species of trees for example, can live for several hundred years, grossly overshadowing the average lifetime of most humans. It is also interesting to note that different classes of individuals within the same species can also have disparate lifespans, such as a queen bee which can live up to 5 years while worker bees, with the same genome as the queen, only live for 40 days. This is also evident in humans, where different cells of the body also have varying lifetimes. For example, epithelial cells of the small intestine live approximately 36 h, cells of the epidermis live for 10 days, liver cells live for 450 days, and so on. Nevertheless, the average lifespan of human beings is more than 70 years.

From multiple analyses of lifespans in different representatives of various species, several conclusions can be drawn: (a) for every species, there is a maximum inherent duration of life which is determined genetically; (b) the duration can differ greatly between closely related species, and even within each species; (c) larger animals generally live longer than smaller ones, though there are many exceptions; (d) there is no clear connection between the duration of an

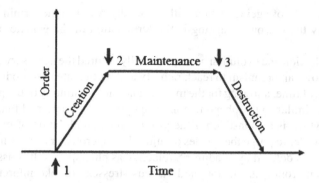

**Fig. 11.2**  The maintenance and destruction of living bodies. *1*—Start of life at the moment of fertilization. Switching on of genetic programs of creation, development, and maintenance. *2*— Switching off of the programs of development. *3*—Disturbances of the programs of maintenance and spontaneous destruction and ageing of the organism

organism's life and its level of organization and complexity, although more developed species tend to live longer; (e) seasonal changes have effects on the lifespan of some animals (such as insects); (f) the larvae of many insects live considerably longer than mature adults; (g) individuals of the same species, but of different sexes and social statuses, can have significantly different lifespans, even though their genomes are the same. It is difficult to explain the above observations, but it seems evident that there exists a stable internal determination of the achievable lifespan within living bodies.

A number of factors provide evidence for the idea that the duration of life of all organisms is genetically determined: (a) lifespan is a permanent and characteristic attribute of species which is inherited; (b) the lifespan of monozygotic twins is very close, and does not always coincide with the lifespan of fraternal siblings; (c) there are known hereditary diseases that cause accelerated ageing, such as progeria, which are connected to the failure of DNA repair; (d) hybrid animal species can have lifespans which differ from either parental species; (e) it is widely held that, without detrimental extenuating factors, the longer a parent can live, the longer their descendants will too.

Because the ageing process is very complex and affects all organizational levels of every living creature, there is no definite ageing gene, or even one that might determine the duration of life. The entire process of ageing is cumulative, involving the coordinated effort of specific genetic programs from multitudes of genes throughout the entire lifespan of an individual, from the moment of inception, right through development, and continuing into maturity until the moment of death. During embryonic and post-embryonic development, there exist processes which are aimed specifically at the creation and maintenance of order in the living system. Upon reaching pubescence, the stabilizing genetic programs of growth and expansion are terminated (Fig. 11.2). After this point, even though the organism possesses an inherent stability and the capacity to regenerate, it begins the natural process of ageing, which will eventually result in the termination of its life. Evidently the reason for ageing is the termination of the genetic program of development.

Natural selection and evolution have genetically secured the necessary duration of life which allows an organism to reach puberty and participate in reproduction for a certain period of time, allowing for the maintenance and continuity of the population. Furthermore, similar to the phenomenon of apoptosis, it is assumed that there is a mechanism which is responsible for the programmed annihilation of multicellular organisms in order to purge the species of injured, dangerous, or useless individuals. Such a process is defined by Vladimir Skulachev as phenoptosis. It is assumed that unfavorable environmental factors and various stresses provoke infarcts, strokes, cancer, and other disorders that act as the *instruments* of phenoptosis, the probability of which increases with age. The concept of phenoptosis may be broadened by adding 'monocarpic' organisms, which reproduce only once and then rapidly die soon after (salmon, eel, annual and biennial plants, etc.). There is also a range of variations of suicide, which is inherent to certain species such as humans.

However, neither the death of cells (the smallest phenotypic framework of genomes), nor the death of an organism (a formed colony of cells), can stop the phenomenon of life, which continues on through the cyclical reincarnation of genomes.

*The lifetimes of species.* Numerous species of organisms live (as a whole) for long periods of time, but not a single one can survive forever, at least not without significant alteration. The average lifetime of a species is on the order of several million years, in spite of the significantly shorter lives of the individuals within it. For example, many modern species of insects such as ants, dragonflies, cockroaches, and so on, have already lived on the Earth for tens of millions of years, and certain crustacea, brachiopoda, and reptiles have lived for hundreds of millions of years! The basis for their prolonged existence is the ability of individuals to transfer stable and practically unmodified genetic information to their descendants through reproduction. In other words, over millions of years and generations, genomes have been transferred from cell to cell and from organism to organism, thus maintaining the continuity of life.

There are many factors which can lead to the death and disappearance of a species, such as planetary catastrophes, evolution of the Earth, changes in the environment or their habitat, or competition. Any of these or a multitude of other events can significantly limit the survival and reproduction abilities of individuals within a given species, causing it to die off faster than it can regenerate, eventually resulting in the termination of its genetic continuity. Throughout the billions of years of Earth's development, considerably more species of living organisms have disappeared than exist today, some of the most notable being dinosaurs, pterodactyls, arborescent horsetails, and others.

It is quite possible that the duration of life of a species is also determined genetically. Indeed, not only can the whole species disappear, but, more importantly, their entire genomes can disappear. A genetic network that is composed of billions or trillions of special and unique genomes can completely vanish! If the programmed death of a species is stipulated by genetics, then this process of elimination can be called genoptosis (by analogy with apoptosis and phenoptosis). But how could this happen and why? In genomes themselves, there are fairly abundant segments with unknown functions. Many of them are remnants or integrations of viruses, bacteria, or mobile genetic elements, which can all be activated at any moment by some unforeseen event, with unpredictable consequences. A further risk to the existence of stable genomes is related to the continuous genetic transmission within the Global Genome, by either a vertical or a horizontal transfer of genes. In other words, there is a constant shuffling of genes within the system of the Global Genome, through which unpredictable and possibly detrimental gene combinations can be created. One potential consequence of such changes is that certain genomic species may become lethal.

The inevitable death of living bodies is caused by the laws of physics. All living systems are destined to die from the point of view of thermodynamics. It is known that living bodies are complex, open, unbalanced, and dissipative systems, whose very existence contradicts the laws of physics which only want to destroy them and restore the molecular chaos from which the body was formed. Nevertheless, bodies can and do exist for a limited period of time by the forceful and energetic influx and utilization of free energy and matter from the external environment, as well as by a targeted use of genetic information as a force of order. The duration of the life of biosystems is therefore determined by the extent and effectiveness of the functions of genetically determined integrity-maintaining processes.

## 11.3 Maintenance of Integrity

Nothing is eternal. Material bodies are constantly under the impact of various stresses from the environment, such as thermal motion of molecules, various types of radiation, fluctuations in temperature, and so on. These stresses can result in injuries to and wear of various body parts and vital systems. If organisms did not have special mechanisms for self-restoration and maintenance of their integrity, they could quickly die under such duress. In order to survive and thrive, organisms must therefore have at least some rudimentary genetic, biochemical, and physiological mechanisms which can counteract negative forces and restore the balance of continuous life. The underlying principle of these various mechanisms of self-preservation is the genetically determined necessity and capacity for constant physiological regeneration at all organizational levels.

- *Constant regeneration of macromolecules and cellular organelles.* After synthesis, most protein molecules exist, on average, for 12–20 h. Worn-out and used proteins are destroyed by proteases to form free amino acids, which are re-used to create new functional proteins. The potency of this regenerative process is exceedingly high and, within several days, the entire protein composition of a human being is practically completely renewed. Similar processes also occur

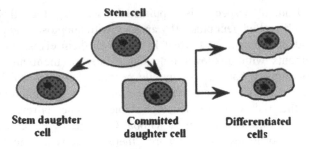

**Fig. 11.3** Stem cells: keepers of the native genome and the primary phenome. These cells represent the powerful potential of development and the maintenance of integrity. They also possess the unique ability of asymmetric division, which allows semi-differentiated progenitor cells and new stem cells to arise from a single division event

with other macromolecules, their complexes, and organelles. The regeneration of DNA, for example, is secured by various reparative processes which act to immediately restore damaged segments with the help of special protein enzymes. Furthermore, the preservation of an unchangeable cellular DNA content is conditioned by periodic processes of replication that occur before the division of the cells, ensuring that each daughter cell will have an identical genetic composition.

- *Regeneration of the cellular constitution.* This mode of regeneration primarily deals with tissues, and is carried out through the replacement of damaged or retired cells by new ones. The ageing of cells can be natural or external, and can result from such factors as physical injuries, harsh environments, or accumulations of toxic substances. Because the maximal lifespan of a cell is determined genetically, naturally occurring destructive processes within cells can be greatly accelerated after stress or damage, resulting in their elimination by the endogenous activity of their own lysosomes, which cause the cells to become fragmented and then devoured by phagocytes. Upon their eradication, new functionally active cells arise to take their place after successive cycles of division and, where applicable, by the differentiation of stem cells (Fig. 11.3). Analogous to protein turnover, the intensity of this process is also quite high, resulting in the complete recycling of a human being, for example, within a period of approximately 7 years.

Although the cells of most organs are relatively short-lived, there do exist some that share the same lifespan as the tissues they comprise. Some of these long-lived cell types include neurons, lens cells, and cardiomyocytes. These cells form either during the embryonic period or at early stages of post-embryonic development. They never divide, and, in case of death, may never be replaced. The mechanisms which allow these cells to live for such long periods involve specific genetic programs which constantly renew enzymes, structural proteins, biological membranes, cell organelles, and other parts of the cells. For example, neurons function for dozens of years, primarily due to the unhindered regeneration of cellular components.

Because most vertebrate cells are not permanent, they are typically replaced by new cells over time. This is generally done in a naturally premeditated manner. However, in some cases where a part of a specific organ suffers massive damage, the remaining cells of the organ attempt to compensate for the loss of function, or increase their rate of division to fill the gap. This is evident in the case of liver cells (hepatocytes): if a significant portion of the liver is destroyed as a result of an injury or intoxication, the division rate of the remaining hepatocytes grows exponentially to repair the damage.

Every tissue and organ consists of many types of cells. The mixture of different cells, which constantly die and are replaced by new ones, is maintained thanks to the cellular memory found within every cell lineage. In other words, a dead cell is replaced with an exact replica, which allows differentiated cells to maintain the structure and functions of organs and their homeostasis, and to transfer properties of specialization to daughter cells. Some tissues, such as the epithelium of the small intestine, skin, and blood, are intensively renewed due to the division of non-differentiated stem cells. Although they are generally present in tissues in small amounts, they serve as the ancestral cells for many future generations of differentiated cells. Stem cells possess the following properties: (a) they are not specialized; (b) they are able to divide without limitation; (c) after division, one daughter cell remains a stem cell while the other one is differentiated into the required type of cell.

Stem cells of certain tissues are genetically determined. For example, muscle cells are formed from skeletal muscle stem cells, sperm is formed form spermatogonia, and the basal cells of the epidermis develop from epithelial cells. These types of stem cells are unipotent, meaning that they can only differentiate into a single cell type. Pluripotent stem cells are significantly less restricted in their development, and therefore have the ability to differentiate into a whole group of cell types. For instance, blood-forming pluripotent cells of the bone marrow can give rise to erythrocytes, leukocytes, monocytes, lymphocytes, etc., the selection of which is connected with the capacity of the cells to selectively express only those genes specific for a given cell type. Totipotent stem cells are the least differentiated of all stem cells, and can establish all the necessary units of the organism. Only embryonic blastomeres possess this property until a specific stage of development.

The integrity of multicellular organisms is also maintained due to reparative regeneration, which is the restoration of damaged or rejected organs or entire parts of the body. For example, it has been shown that children who lose fingertips in accidents may re-grow them in just a few months provided that there are no other complications. Significant parts of bodies can be shed and later regenerated in other creatures as well, including *Annelida* worms, starfish, and some mollusks. Some arthropods, amphibians, and reptiles can regenerate limbs, and some species of lizards can completely break off their tails in the event of an attack, while these then grow back quickly and completely. The healing of dermal tissues, bone fractures, or injuries of internal organs can also serve as examples of reparative regeneration. Some species of plants can even develop new organisms from somatic cells. This process of restoration is called somatic embryogenesis, since it very closely resembles much of the development of an embryo.

The maintenance of integrity and order is impossible without energy expenditure, and it may require the use of as much as 90 % of the total energy produced by an organism. An equally important condition for the maintenance of integrity is the use of genetic information which controls the formation of all the necessary components of complex biological systems. It is apparent that the main basis for the continuance of the life of organisms, as well as the support of their homeostasis and order, is the constantly controlled regeneration of the molecular structures within cells and cellular compositions of tissues and organs, as well as various processes of renovation and repair. In other words, living bodies possess tremendous internal dynamics regarding the exchange of their components, and this allows cells, tissues, and organs to maintain a certain level of constancy in their composition, volume, and form. This characteristic can also be extended to species and the organisms that belong to them. Species have significantly longer lifetimes than their individual representatives, thanks to the constant transfer of genomes to future generations as older and weaker members die off. In this way, a biological species can live on Earth for millions of years with a relative preservation of genomic cohesiveness and increasing adaptability through natural selection.

In conclusion, living systems are constantly renewed at all levels of organization. Their ability to regenerate is closely connected with the main characteristic of all living bodies—reproduction. All processes involved in the maintenance of integrity exist for one specific objective: genomically controlled maintenance of its own homeostasis through renewal of its phenotypic framework.

# Chapter 12
# Ageing and Death of Individuals

## 12.1 Ageing

This is the gradual process of decreasing vitality in organisms, connected with the action of unfavorable factors which lessen the effectiveness of their functionality.

As noted earlier, the reason for the inevitable ageing and death of living organisms lies in the laws of Nature, which work to destroy the order the organism works so hard to maintain. Living organisms survive for a limited period of time because, as they grow older, they lose their ability to exist in equilibrium with the environment. During the intensive growth and development phases of multicellular organisms, ageing is virtually not manifested, but it usually appears immediately after pubescence. Afterwards, with passing time, ageing processes increase and intensify at all levels of organization, from single molecules to entire body systems.

*Ageing of Cells.* Cellular ageing is rooted in molecular events. The majority of all functional molecules are altered over time in the aqueous medium of the cytoplasm, primarily as a result of interactions with other molecules and atoms (thermal motion, chemical reactions, free radicals) and through changes caused by radiation. Molecules can decay into atoms, transform into other molecules, undergo structural changes, and denature. Therefore, such changes can significantly impact the efficiency of key functions in biological systems.

One of the main damaging factors at the molecular level is the presence of free oxygen radicals in living cells. These are highly reactive molecules with an unpaired energetic electron. They are formed as by-products in processes of energy transformation in the respiratory cycle (electron transport chain) of mitochondria, as well as a number of other metabolic reactions. These molecules are highly reactive, and can attack and damage various organic molecules, including nucleic acids and proteins. They can be blamed for such illnesses as cancer, diseases of the heart and blood vessels, and other disorders that accompany ageing.

These and a number of other damaging effects may cause the oxidation of lipid molecules in cell membranes, inactivation of enzymes, glycosylation of structural

G. Zhegunov, *The Dual Nature of Life*, The Frontiers Collection,
DOI: 10.1007/978-3-642-30394-4_12, © Springer-Verlag Berlin Heidelberg 2012

proteins and the formation of cross-links between them, damage to DNA, mutation of genes, etc. This in turn leads to a gradual destruction of the cellular structure and a deterioration of its functions. The integrity and barrier functions of membranes degrade, enzymatic activity decreases, the cell becomes littered with products of metabolism, and the synthesis of proteins and the regulation of cellular processes cease. All these effects result in a catastrophic disturbance of the regulation of cell functions, leading to the emergence of systemic "diseases of ageing" which weaken an organism's resistance to external stresses, eventually leading to death.

Cells have certain "anti-ageing" mechanisms which allow them to cope with damaging environmental factors, at least for a while. For example, the very ancient enzyme superoxide dismutase transforms oxygen radicals into hydrogen peroxide, which is subsequently broken down into water and oxygen by catalase. Other examples include enzymes that restore damaged segments of nucleic acid molecules (nuclease, polymerase, and ligase), break up damaged proteins (proteinase and peptidase), and restore denatured proteins after the action of destructive factors (chaperons). These mechanisms significantly slow down the processes by which cellular structures wear out, as well as the actual ageing process, but they cannot completely suppress it.

It is important to note that ageing is a very complex phenomenon involving many interconnected processes. Thus there is probably no special gene for ageing, and ageing processes may depend on the action and regulation of several hundred genes. This means that understanding and influencing the role of just one of them (e.g., the gene telomerase) is not a viable means of solving the problem of ageing.

The molecular and cellular features of ageing are multifaceted. In somatic cells of animals, ageing is accompanied by an overall decrease in the intensity of RNA synthesis. For example, up to half of all rRNA genes deactivate in the ageing organism, and certain mRNAs disappear while others appear. Also, as cells and organisms get older, the content of non-histone proteins in chromatin also decreases, and their linkages with DNA become less labile. Thus, there is a partial alteration of the genetic information used by a cell at various periods of existence. This leads to the fact that, with age, the qualitative and quantitative set of proteins in a cell also changes. This particularly concerns the various enzymes responsible for anabolic processes and the maintenance of integrity. The alteration of gene expression through somatic mutations can also expedite the ageing process. Changes to the structure of genes can lead to problems in the synthesis of various proteins or decreasing abilities of protein-synthesizing systems. Depending on the role of the affected proteins, vital systems such as metabolic and adaptive systems may be compromised, limiting the potential for cells to repair and function. For example, the activity of enzymes responsible for coping with oxidative stresses decreases in the ageing organism, leading to deteriorations in the transformation of energy and a decrease in the number of cellular mitochondria.

Ageing affects practically all organelles, general and specific alike. One of the most noticeable age-related changes is the rearrangement of permanent highly specialized cells, such as neurons and cardiomyocytes. For example, ageing nerve cells display a characteristic redistribution of membrane structures to the cyto-plasm, a decrease in the volume of the rough endoplasmic reticulum, and an increase in the content of microfibrilles, which may disturb the transport of sub-stances along the axon. An accumulation of unusual substances such as lipofuscin can also occur, clogging up the cell and altering its functions. As a result, there is a decrease in the conduction of nerve impulses, and, in some types of nerve cells, there is also a decrease in the amount of neurotransmitters.

*Ageing of multicellular organisms.* Since natural processes obey the laws of thermodynamics, spontaneous destructive events are favored because they increase net entropy, while creation requires an expenditure of energy. Therefore, Nature strives for a constant increase of chaos through the dissipation of matter and energy. In some sense, living organisms resist Nature and even generate more and more complex structures, accumulating information in the form of order. Because these processes are in direct opposition to Nature's demands, ageing is the process by which all life forms are brought back into compliance. For example, the order and complexity of an embryo increases abruptly immediately after fertilization. The overall number of cells quickly increases, and these cells then differentiate and separate in specific ways, whereupon tissues, organs, and other parts of the organism begin to take shape (Fig. 12.1a).

This creation of order occurs due to a unique program of genetic development, which then switches over to one controlling the maintenance of structural and functional complexity during the lifetime of the organism. The highest level of order is attained by the time the organism reaches sexual maturity. Upon reaching this stage, the development program comes to an end and the programs for the "support of integrity and the struggle against entropy" are initiated. Thus, Nature leaves an organism to the mercy of fate, whereupon the level of order starts decreasing grad-ually under the action of thermal molecular motion and unfavorable factors of both the internal and external environments, accounting for the process of the ageing of cells, tissues, and organs. As ageing progresses, the ability of an organism to reproduce also diminishes, and then finally terminates. After death, natural processes of disorder are fully activated, ending with the ultimate dissipation of the organism at the molecular level. Thus, there is probably no specific genetic program of ageing.

Living bodies can successfully combat entropy only because of the existence of non-obsolescing genetic programs of development, which are repeatedly rewritten and transmitted to succeeding generations through the mechanisms of DNA rep-lication and reproductive processes. The development of the embryo and the birth of the organism together with its subsequent growth, reproduction, ageing, and death are all regulated by a system of genes that interact with each other. A living organism can be compared to a book that is constantly reprinted. The paper on which the book is written may be worn out and decayed, but its content is eternal if the matrix of its message (the program of its creation) is preserved.

Thus, differential gene expression is observed in the process of ontogenesis of multicellular organisms. Different systems of interacting genes operate at different stages of embryonic and post-embryonic development, assuring the gradual, pre-cise, and stepwise development of an organism. Over the lifetime of an organism,

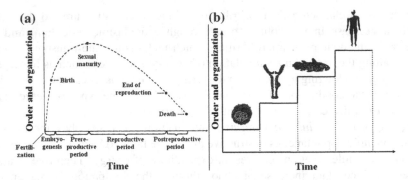

**Fig. 12.1** Order and organization of living systems in the process of development. **a** Thermodynamic trajectory of the progression of living bodies throughout the life cycle, differing in principle from the constant step-by-step complication of the Integrated Life System, **b** It is obvious that the internal energy and negentropy of the integrated system of life increase in the process of evolution. However, this is not the case for an individual living organism, since it increases its negentropy only at the initial stage of development, whereupon this process stops and the body starts ageing and undergoing gradual destruction, i.e., the entropy increases and the internal energy decreases

different gene networks are established which support its integrity, including the special groups of genes that aid in the differentiation of cells and creation of tissues. In other words, it may be concluded that the life cycle of an organism is a cycle of the successive, well-ordered, and differential expression of various systems of genes of a permanent genome. Thus, the origin of the life of an individual, its development, and its survival are rooted in the totality of transformations of matter and energy based on the use of genetic information and in cooperation with the external environment. From this perspective, we can say that one of the primary reasons for the existence of ageing and death in living organisms may be disturbances in the differential expression of gene systems.

As long as the formation of life and its evolution took place under conditions of limited resources, it was advantageous for Nature to destroy concrete individuals once they had fulfilled their reproductive role by down-regulating the set of genes that manage the processes of development and regeneration. This decrease in vital activity occurs in different individuals at different times, and this determines the apparent interspecies and intraspecies variability in longevity.

The hypothesis of telomerase ageing is also associated with damage to one of the regenerating mechanisms which deals with the maintenance of integrity. The termination of the functioning of telomerase is connected with the cessation of the genetic program which controls its production. The function of telomerase is connected with the regulation of the number of cell divisions (that is, the lifespan of a population of cells—*Hayflick limit*, 1961), but not with the struggle against the molecular ageing of cells themselves. Thus, telomerase deregulation may be just one of the reasons for the ageing of a multicellular organism, but not for the ageing of individual cells.

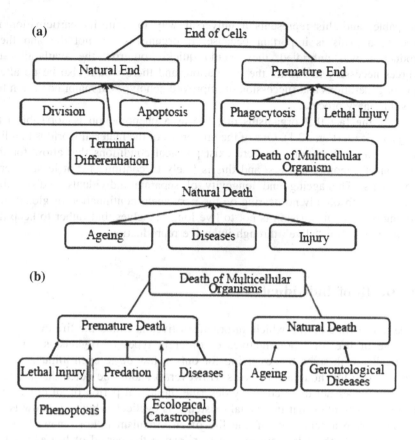

**Fig. 12.2** Role of ageing and death in the downfall processes of living bodies. **a**—chart of the main reasons for cellular downfall. **b**—chart of the main reasons for a multicellular organism's downfall. The majority of cells and organisms die not from ageing, but for various other natural reasons

It is important to consider the ageing (and death) of a multicellular organism as a colony of cells with a single genome to be a completely different process from the ageing and death of the separate members of the colony. Cells are individual autonomous micro-heterogeneous systems, while multicellular organisms are macro-heterogeneous systems constructed from a large number of these systems. If for separate cells the decisive factors in ageing and death are thermal motion, radicals, and other molecular processes, then the reason for the ageing and death of a multicellular organism is the accumulation of damage in the systems of maintenance and regulation of cell colonies, organism parts, organs, and other functional systems. Meanwhile, the rapidity of ageing and the lifetimes of cells, as well as organisms, are determined by the effectiveness and continuation of regenerative mechanisms.

Overall, ageing leads to a progressive increase in the probability of death. The biological essence of ageing lies in the fact that it makes the death of an organism inescapable, and this represents a universal way to limit its participation in reproduction. This is important because, if organisms did not die and their reproduction was unbridled, they would quickly outgrow the availability of resources necessary to sustain the population, and there would also be no alternation of generations and succession of improved genotypes, a major condition for the evolutionary process.

Despite the fact that organisms die, life as a phenomenon is not subject to ageing and death (Fig. 12.1). One of the properties of the material world is that life will exist on Earth as long as there exist physical conditions that allow for the habitation of genomic carriers, and this is likely to continue for at least several billion years. The ageing and longevity of separate individuals—"disposable somas"—has absolutely no bearing on the permanent continuation of global life. As living beings, our mission is not to live long ourselves, but rather to keep the phenomenon of life "alive" through intensive reproduction.

## 12.2  Death of Individuals

The period of time during which an organism lives is called the life cycle (LC). The cycle of life for representatives of different types of organisms can vary dramatically, with some bodies living for only a few days, even minutes, while others can live for thousands of years. However, it should be noted that the term "life cycle" does not quite correctly describe the actual period of an organism's life because it implies that individual existence is cyclical, which we know is not true, since from a certain point, the life of an organism is not renewed, but is eliminated by death. It is more logical to perceive the period of life as a set of processes that allow for survival, duplication, and spreading of specific genomes. Nevertheless, it is true that a "cyclic" recurrence is characteristic for genomes, since they are reproduced in a multicellular organism both through internal cellular division and by reproduction. Even after the death of an organism, its genome continues to exist in its descendants, establishing a cyclical renewal of the genome in future generations.

Let us now consider the role of ageing and death as a global process of overall destruction. According to the laws of physics, the downfall is imminent for every natural object, including living bodies. Downfall is a general term that means an act of ending something's existence. Death is a particular case of downfall that is appurtenant to living bodies. It is the ultimate ending of an organism's physical life due to the termination of the main processes and functions that maintain it. Death is preceded by the process of dying, which is connected with a cessation of the main processes and functions. After death, a lifeless body usually remains, and this is gradually destroyed over time until no trace of the former carrier of life is left.

Certain peculiarities in the processes of downfall of individual cells and multicellular organisms should be noted (Fig. 12.2).

A. The natural downfall of individual cells can hardly be considered a death, since it occurs as a result of pre-determined cytogenetic processes such as division, apoptosis, and terminal differentiation. In these cases, the obligatory process of the natural death of individuals is absent. Phagocytosis and lethal damage may lead to premature downfall. A mass downfall of cells occurs after the death of multicellular organisms, whereby many cells perish without development of an ageing process, and at much earlier stages than the maximal possible length of their life. Natural downfall is conditioned by internal reasons, while premature downfall is conditioned by external reasons. The natural death of cells is associated with ageing and with nonlethal damage or diseases. These lead to irreversible deterioration of structures and unbalanced functional bodily systems. The cell dies gradually, and eventually completely ends its existence.

B. Multicellular organisms are also prone to premature downfall. This occurs primarily for external reasons such as carnivore attacks, severe diseases, lethal damage, etc. The death of monocarpic animals (mayflies, humpbacks, calamari, etc.) occurs suddenly, right after reproduction. Such a phenomenon may be considered as phenoptosis (defined above), which occurs as a result of internal genetic controls. A mass premature downfall of organisms can also occur, and may generally be attributed to ecological catastrophes (forest fire, flood, impact of large meteors). Thus, complex organisms may also perish prematurely, without showing any aspects of ageing. On the other hand, natural death is associated with reaching or coming close to an organism's maximal age, and may occur as a result of ageing and chronic diseases. It is apparent that ageing and death are just a small part of the global process of downfall. It is only for human beings that ageing has a particular role, since we are the only organisms that can rationally and knowingly protect themselves from unfavorable factors.

Death is a hereditary phenomenon in the life of each organism and is a natural event which occurs after a specific period, provided that it does not occur prematurely due to other external factors. The maximum duration of life, as a feature of a given species, is passed on as a heritable trait to future generations, and explains why trillions of individuals from billions of generations of different species have lived, live, and will continue to live within distinctly defined time frames. There are multiple examples which show that death is a genetically predetermined condition. For example, mayflies live only one day. During this period they have to mate and lay eggs, after which they all die. Thus, in this case, death is conditioned not by ageing, but by phenoptosis—sudden hormone-determined self-annihilation. The imagoes of many insects live only for a few days, while their larvae are capable of living for years. Other examples include males of some squid species that die immediately after insemination of females, and annual plants that live for only one season. There are many other facts and cases which show not only that life is program-based, but also that individual death is as well. Nevertheless, for a system of genomes within a species, the longevity of individuals is of virtually no importance. The only facet of import is their ability to successfully reproduce.

Although the death of an individual is an unfortunate and often tragic event, it is insignificant as far as its genome is concerned and is actually generally positive for species and communities. The death of an old organism provides a place for a new, young, and healthy individual that may have new combinations of features that increase its ability to survive, adapt, reproduce, and evolve. The processes of death and birth through reproduction maintain an optimal and balanced number of individuals within a population. Thus, the existence, survival, and propagation of species are controlled by the processes of birth and death, and these in turn are controlled by DNA programs. We should also note that the concept of a well-defined lifespan exists both in complex multicellular organisms and in simple unicellular ones, which are often found in colonies containing old, young, dividing, dying, and dead cells.

The remains of dead cells or their derivatives play an important role in the lives of some organisms. For example, the dead cells that form sclerenchyma in plants provide support and durability even after their death. Furthermore, the wood of trees contains vessels which ensure the flow of water and dissolved substances throughout the entire tree, and are also formed by mortified cells. As a matter of fact, hair, claws, horns, hoofs, and feathers are all structures that are composed of dead cells and play a significant role in the lives of the organisms in which they are found. Even the outer stratum of animal skin consists of several layers of dead cells which help defend an organism against loss of water and protect it from the penetration of microorganisms.

Genetically programmed cell death (apoptosis) is also a natural occurrence during the development of many organisms, and occurs at a specific time and place. For example, certain cells of the budding hands and feet in human fetuses undergo apoptosis to allow the digits to separate, thus forming fingers and toes. Furthermore, bodily cavities, vessels, and organs are all formed from a similar process of cell death.

It is therefore apparent that during the evolutionary process, there arose hereditary mechanisms and mechanisms of natural selection which determined the lifespans of individuals from all species of organisms at the cellular and molecular level. This, in turn, became beneficial for communities of such organisms, because individual interests were sacrificed for the sake of the prosperity and continuity of a community or population containing a certain genome.

Therefore, after the death of individuals, life as a phenomenon does not stop, as long there are new offspring which live on to propagate the genome of the species. The substrate of life is the set of DNA which is transferred in the process of replication (cellular) and reproduction (organismal), and which continues to live on in the new organisms. Bodies, therefore, are only an instrument of existence, replication, and dissemination of a permanent genome. As the life cycle continues through the production of new individuals, slight genomic changes occur which are passed on to the new generations, maintaining the genetic continuity of life. Thus, as the body of a single discrete carrier of a unit of life dies and decays, the existence of its genome lives on in its offspring, a cycle that can continue for thousands and millions of years and generations.

Therefore, all cells and individuals of all species are mortal, and this is essential to the phenomenon of life and its evolution. Nevertheless, despite the death of individuals which takes place according to the laws of Nature, life as a phenomenon continues, and the existence of species is continuous for many millions of years, thanks to the processes ensuring reproduction of individuals, i.e., the transfer of permanent genomes to new generations. Therefore, after the death of an individual, only the phenotypic portion of life ceases its existence, while the genotypic part perdures in its offspring. Although the phenome is dead, long live the immortal genome and its new phenome!

## Outlines of Duality 2

Regarding the second part of the book, we can draw the following conclusions emphasizing the various aspects of the duality of life:
*Living body vs. the phenomenon of life.*

1. Living bodies are autonomous, but the same cannot be said about the Integrated Life System (ILS), a network that goes right round the surface of the planet. Living bodies consist of cells, and the ILS consists of genomes in their phenotypic framework.
2. The ability to survive is an inherent trait of living bodies, but not a continuous property of life. Living bodies are constantly destroyed and perish, but new ones emerge in their place to maintain the system of the Global Phenome. The ILS appeared many years ago and it does not age, die, or reappear anew, but exists as processes of permanent development in the form of continuous genomes that "travel" from one mortal body to another.
3. Individual living bodies die, and entire species of genomes perish, but the Global Genomeand the phenomenon of life itself does not perish. The death of a separate individual is an inevitable part of the immortality of genomic life through continuous regeneration of the elements of the Integrated Life System.
4. The term of existence is defined for all living bodies, but not for the phenomenon of global life. There is no known immortal organism, although the Integrated Life System has existed continuously since its appearance billions of years ago.
5. Living bodies possess a finished form and content, which may be changed or updated in new offspring through the continuous processes of life's evolution.
6. Reproduction is the primary goal of living bodies, but not for the phenomenon of life itself, though the eternity of the Integrated Life System is achieved by a continuous self-replication of its elements.
7. Elements of adaptation, variation, and evolution occur in the reproductive processes of organisms, which serve as the basis for the global evolution of the Integrated Life System. Living bodies are simply a means of survival and reproduction of a unique genome which is a member of the system of the

Global Genome in the ILS. The purpose of reproduction and evolution, therefore, is not the survival of the organisms and species, but the survival of the genomes and the information which they transfer.

8. The difference between the phenomenon of life and a living body may be illustrated by the following example. Imagine a settlement of humans from Earth on Mars. Can we assume that life has emerged on Mars? No, because humans emerged on Earth, and are therefore not in any way developmentally related to the conditions prevailing on Mars, as can be observed by the mere fact that, without their protective suits, they could not survive on the planet. A system of living organisms that propagate without assistance, survive, and complement the new planet's own nature is not thereby formed. If the settlement is temporary, then after the humans leave, there will no longer be traces of life on Mars, although the settlers will continue to live on as they journey back to Earth. We can therefore say that the presence of humans on Mars is merely the attendance of living bodies that are the representatives and constituents of Earth's life phenomenon. In other words, the establishment of the phenomenon of life is tied to the planet as the life of an organism is tied to its body.

9. The integrating element of the life phenomenon and living bodies is the genome. It combines both into the Integrated System of Life.

*Multicellular bodies vs. individual cells*:

1. Individual cells and multicellular organisms are quite different living bodies, though both represent highly organized, heterogeneous, open, and unbalanced systems.

2. Cells are autonomous systems of selectively chosen molecules and their complexes, while multicellular organisms represent an integrated system that consists of conceptually different cellular elements. In multicellular bodies, single cells are used as building blocks in order to establish higher levels of histological and anatomical structures, as well as functional systems. Therefore, on the one hand, multicellular bodies are built in a simpler way and are just derivatives of interacting cells, while on the other hand, they are more intricate, since they unite and coordinate several levels of complexity.

3. In cells, controlled biochemical and biophysical processes constantly take place, and are primarily directed at the conversion and targeted use of matter and energy to support their own high level of structural–functional organization. Likewise, the numerous organs and tissues of multicellular organisms carry out global functions which are connected with the survival of the entire organism. The micro and macro homeostatic and survival processes are reciprocal, because each depends on the other.

4. The instrument that arranges and controls the structure and functions of cells is the genome. The genome builds up a system of protection and support around itself in the form of the intra- and extra-cellular matrix in and around cells, and this serves the self-preservation, reproduction, and dissemination of the cells. At the organismal level, the instruments that arrange and control the body's

structure and functions are the neuro-endocrine and immune systems that serve to integrate cells, tissues, and organs into a single organism.

5. Based on the understanding that cells are above all containers of a genome, we can suggest that multicellular organisms can be considered as colonies of cells that incubate genetic material. These cells work together to allow the mutually beneficial processes of life, survival, and reproduction to occur by acting in accordance with each other while complying with the demands of their genetic apparatus and the regulatory systems of the organism. In doing so, they maintain integrity and ensure a multitude of functions not only in themselves, but in the entire body. Such organisms are therefore immensely complex communities of the smallest of living bodies, and represent a qualitatively new state of existence of cells and their genomes.

6. Since cells and multicellular organisms are absolutely different biological systems, the mechanisms of their reproduction, development, support of integrity, ageing, and death are also essentially different. Cells proliferate simply by division, while multicellular organisms do so through complex developmental and reproductive processes resulting from the formation and amalgamation of gametes (in case of sexual type of reproduction). In addition, the development of the majority of unicellular organisms is rather simple, while it is much more complex and prolonged in multicellular organisms. The support of integrity and homeostasis of unicellular organisms is based on molecular mechanisms, while multicellular ones rely on various cytological and physiological processes. Lastly, the ageing of unicellular organisms is caused by internal molecular processes, while in multicellular organisms it is generally connected with a disorder of regulation, direction, and coordination of cells, tissues, organs, and their systems.

*Generative cells vs. somatic cells*

1. Multicellular organisms consist of two completely different types of cells: germ line and somatic.

2. Peculiar generative cells already begin to specialize at the early stages of embryonic development in a multicellular organism. They migrate within an embryo and gather in a certain location. These germ line cells are the ones that can later create new organisms through fertilization and embryogenesis. They are formed from the gametogenic epithelium of the gonads, which takes part in the production of gametes. Generative cells are unique because they are the only ones that produce gametes - the transitive forms of genomes. Only this group of cells provides continuity for the life process of the given organismal species, or rather, the continuity of a specific genome type. The germ line cells of all species comprise the instrument and the mechanism that serves to maintain the continuity of multicellular life as a global phenomenon.

3. Diversified somatic cells form the bodies of organisms. Although bodies are made up of somatic cells, they merely serve to create an organized environment for the gametes to be produced and then passed on to future generations. In this

respect, somatic cells are of secondary importance to germinal cells, and are merely their servants.

4. A single uniting element of both cellular systems is the genome which does not differ in structure and information between the two types of cells. The difference lies in the phenotypic framework and different genomic determination.

# Recommended Literature

1. Darwin, C.: On the Origin of Species: A Facsimile of the First Edition. Harvard University Press, Cambridge, Mass, (1975)
2. Mendel, G.: Experiments in Plant Hybridisation. Harvard University Press, Cambridge, Mass (1965)
3. Alberts, B., Bray, D. et al.: Molecular Biology of the Cell, Garland Science, New York (1994)
4. Hopson, J.L., Wessels, N.K.: Essentials in Biology. McGraw-Hill Publishing Company, New York (1990)
5. Cavalier-Smith, T.: The origin of cells: A symbiosis between genes, catalysts, and membranes. Cold Spring Harbor Symp. Quant. Biol. 52, 805–824 (1987)
6. Li, W.-H., Graur, D.: Fundamentals of Molecular Evolution. Sinauer Associates, Sunderland, MA (1991)
7. Levinton, J.S.: The big bang of animal evolution. Sci. Am. 267(4), 84–91 (1992)
8. Becker, W.M., Deamer, D.W.: The World of the Cell, 2nd edn. Benjamin-Cummings, Redwood City, CA (1991)
9. Gilbert, S.: Developmental Biology. Sinauer Associates, Inc. (1988)
10. Raff, R.A., Kaufman T.C.: Embryos, Genes, and Evolution. Macmillan, New York (1983)
11. Margulis, L.: Origins of Eukaryotic Cells. Yale University Press, New Haven (1970)
12. John, P.C.L.: The Cell Cycle. Cambridge University Press, New York (1981)
13. Medawar, P.B.: An unsolved problem of biology. H.C. Lewis and Co Ltd, London (1952)
14. Joyce, G.F.: Directed molecular evolution. Sci. Am. 267(6), 90–97 (1992)
15. Timofeeff-Ressovsky, N.W.: Genetik und Evolution Z. Ind. Abst. Vererbl.–1939.–Bd 76, # 1–2.–S. pp. 158–218
16. Weismann, A.: Das Keimplasma. Eine Theorie der Vererbung, Jena, (1892)
17. Weismann, A.: Vorgänge über Deszendenztheorie, 3 Aufl., Jena, (1913)
18. Skulachev, V.P.: Mitochondrial physiology and pathology, concepts of programmed death of organelles, cells and organisms. Mol. Aspects Med. 20, 139–146 (1999)
19. Singer, S.J., Nicolson G.: The fluid-mosaic model of the structure of cell membranes. Science 175, 720–731 (1972)
20. Cohn, J.P.: The molecular biology of aging. Bioscience 37, 99–102 (1987)

# Part III
# How Life Works: Mechanisms and Processes of Living Bodies

# Chapter 13
# Bodies, Processes, Mechanisms and Interactions

## 13.1 Controlled Interactions

*Living bodies* are autonomous, physical, and highly organized biological systems of interacting molecules that possess biological characteristics and properties. Their various interacting processes are determined by certain mechanisms.

A *process* is a sequential change of conditions and developmental stages in something over time. For example, the process of cell division is defined by a series of sequential events that change throughout mitosis, ending with cytokinesis.

A *mechanism* is a device or mode that activates and determines a phased order of systemic activity. For example, one of the main mechanisms of glucose metabolism is glycolysis.

All natural processes and mechanisms are conditioned by physical and chemical interactions. An *interaction* is a perceived or physical contact between bodies or particles that leads to changes of their *states*. During such reactions, energy may be transformed, as in the case of physical interactions, or both energy and matter may be transformed, as in the case of chemical interactions. The resulting changes in the states of the interacting molecules (changes in their physicochemical properties) leads to a modification of the biological system. In living systems, the majority of interactions are controlled and regulated by various means, to ensure efficient control of the qualitative condition of any part of a cell or organism. Different states can iteratively and reversibly arise within such systems, and this may result in the establishment of some process or function (Fig. 13.1).

The processes of interaction and consolidation are the main sources of essential new properties and the further evolution of living bodies at all levels of organization. For instance, naturally determined interactions and consolidations of certain molecules form macromolecules of defined structures, biomembranes, enzyme complexes, and other systems that possess qualitatively new properties compared with the units from which they are formed. The consolidation of various organelles

G. Zhegunov, *The Dual Nature of Life*, The Frontiers Collection,
DOI: 10.1007/978-3-642-30394-4_13, © Springer-Verlag Berlin Heidelberg 2012

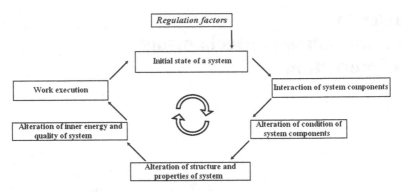

**Fig. 13.1** Interaction of elements within biosystems is the main mechanism underpinning their existence and activity. The regulated interaction of components changes their state. This leads to modifications in the structure of the system as a whole, resulting in changes in its inner energy. Such changes may in turn ensure the performance of a certain task, or condition a new state of the system and a new level of interaction between its elements

and their coordinated functioning conditions a qualitatively new cellular level in the organization of life. In their turn, cells, also differing in their unique structures and functions, consolidate to form an even higher level of organization of life—functioning units of tissues and organs. Thus, a biological system of any complexity always possesses new properties in comparison with units from which it is derived, and this is highly important for development and evolution. At a certain level in the organization of living bodies, new *functions* emerge based on the various properties and processes of their constituents. A function is a specific controlled activity of a certain biological system (cells, tissues, and organs) directed towards the maintenance and retention of life.

It might appear that living bodies and processes are completely different categories, reflecting independent properties of Nature. Bodies, whether single-celled or multicellular, are autonomous and stable formations, while metabolic processes and their mechanisms (protein synthesis, glycolysis, etc.) involve the intracellular dynamics of interacting molecules. However, despite the substantial perceptible differences between separate bodies and their processes, they are essentially quite similar and integrated in many ways. Both bodies and processes are united by their common units of structure and function—molecules. For bodies, they form stable ordered interacting complexes and for processes, they participate in flexible interacting and transformative systems. Bodies and processes only differ in the duration and scale of their molecular interactions and transformations. Molecules that determine processes are extremely dynamic and prone to significant conversion. Molecules that form stationary complexes are less dynamic and undergo conversions at lower rates and only within the determined limits of existence of the given system.

Thus, every structure of a living body is the peculiar result of molecular interactions and processes, resulting in a form of existence where the inner molecular bonds are stronger than impacting outer forces (in particular, the thermal energy of various environmental molecules). Let us look briefly at biological membranes, which consist of a system of phospholipids and proteins, as an example of such stability. The different forces keeping the membrane molecules together, namely hydrophobic, hydrogen, polar, and covalent interactions, are much stronger than the impacting forces from the constant bombardment of the membrane by molecules of water and metabolites. Although molecular processes and metabolic mechanisms also often involve the formation of certain structures and complexes, their lifetimes are generally very short, averaging just a few microseconds.

The difference in the scale of spatiotemporal interactions and transformations of molecules within bodies and processes creates an artificial impression that there may be fundamental differences between them, but this is not necessarily true. In reality, both possess the same molecular nature, and both are the result of interactions, which lead to further modifications of one or the other system. We may say that interactions are the reason for the emergence of structures, which become the reason for higher-level interactions, and so on.

Complexes of physical bodies form biological systems. These "biosystems" are very dynamic formations, because they consist of countless interacting elements. Biosystems typically possess discreteness and several levels of organization. Because the representatives of each organizational level have their own spatial–temporal characteristics of variability, it becomes even harder to grasp the unity between bodies and processes. Moreover, the same molecules can be units of fixed structures while at the same time participating in various processes. For example, carboxylic acids, which serve as constituents of phospholipids, are present in all kinds of very stable biological membranes. At the same time, they are also one of the main substrates of biological oxidation and the transformation of energy. Thus, identical molecules may in some cases be required for stability, while in others they are used up with very high rates of variability.

A cell is also an individual integrated structure. If we observe it as an autonomous physical body, we can notice completely different rates of variability. For example, the lifetime of a cell is considerably longer than that of its components. The entire process of cell division is much longer compared to the instantaneous processes of atomic and molecular transformations that occur during division. The interaction of cells with one another is a rather strong and continuous cooperation of separate large-scale physical bodies. Thus, the variability within an integrated cell is much slower and occurs on a much greater spatial scale that that of the single molecules making up the cell. Cell dynamics is manifested in numerous cytogenetic and physiological processes (e.g., mitosis, motion of organelles and cellular parts, assembly and disassembly of the cytoskeleton, transportation of substances, etc.). Even so, the cell remains an individual integrated body. Moreover, it is important to note that all the cellular macro- and microstructures and their interactions and dynamics are conditioned by molecular-informational processes. In particular, this includes the transcription and synthesis of structural and regulatory molecules, synthesis of the required enzymes, and realization of coordinated chemical and physical interactions.

The state of every biological system is determined by its structure and internal energy. The state of such a system is a function of entropy—a measure of energy dissipation. Entropy characterizes the direction of heat-exchange processes between the system and the environment and within the closed system. Any variation in the internal energy may be accompanied by a change in the state of the given biosystem. Similar transitions may also result from spontaneous or forced modifications of structure or from changes in its qualitative or quantitative composition. Such rearrangements of the structure or internal energy may occur under the influence of various factors of the internal or external environment. For example, they may be caused by such things as the impact of various types of radiation, heat, or chemically active molecules. That is, a minimal signal may cause dramatic changes in the condition and internal energy of a biosystem, entailing the implementation of a certain amount of expedient work.

Changes in the internal energy of bodies or biological systems are generally accompanied by structural adjustments. Likewise, structural changes may be accompanied by changes in the energy of a system. Such rearrangements condition the alteration of the physicochemical properties of living bodies or systems in what are called state transitions. That is, the same system may possess different properties in different states (Fig. 13.1). For example, molecules in cells may be activated or deactivated, oxidized or reduced, exist in various conformations, form monomers or polymers, be charged or neutral, etc. The state of water in cells may be liquid, solid, liquid-crystalline, bound, etc. Even the state of the cellular cytosol can be different and possess different properties, e.g., gel or sol, liquid or meso-morphic, with or without phase boundary, etc.

It is by such transitional adaptability that cells and organisms exist as highly labile and constantly dynamic systems. Through this disequilibrium, biological systems possess a capacity for controlled transitions from one state to another, allowing the performance of correspondingly different modes of work and functions, after which the system may make a transition to yet another state or return to its initial one.

Thus, on the one hand, life is a set of specific processes which organize matter, and on the other hand, the processes themselves are the result of the organized matter. Living bodies, different processes, and mechanisms are all tightly interrelated and have a common nature which is derived from the interaction of molecules.

When talking about the origin of life, we assume it to be associated with the emergence of living bodies, but it would be more correct to state that it involved the interrelated emergence of bodies together with their processes, which were based on selective molecular interactions. Moreover, it is better to talk about the process of existence of living organisms, rather than just the existence of organisms themselves. Therefore, life is a complex of structures, interactions, processes, and states, which are fundamentally similar for all living beings. Depending on the species of the organism, the structures, processes, and mechanisms may vary, and this is a key determinant in the multiform diversity of life.

# Chapter 14
# Strategy of Biological Catalysis

## 14.1 Selective Biological Catalysis

Living bodies may exist in a very narrow temperature range at the very lowest end of Nature's temperature scale. Under such conditions, many chemical reactions are either impossible or occur very slowly. That is why all biochemical and physiological processes in organisms are realized with the accompaniment of biological catalysts—enzymes. Enzymes are proteins that increase the rates of biochemical reactions many thousand-fold, while themselves remaining unchanged and unconsumed, since they only undergo reversible modifications. The emergence of biocatalysts was a revolutionary event, since they conditioned the appearance of rapid and organized transformations of substances in a cold sea of undisturbed chemical chaos, which would have been impossible under any other conditions. The possibility of selective manifestations of planetary properties arose, increasing the probability of reaction flow, which would have been improbable under these conditions without enzymes. This was also the basis for the formation of the organized closed space and orderliness of cells.

Cells are saturated with enzymes. Several thousand types are easily distinguishable and each type may be represented by millions of copies. The quantitative and qualitative composition of enzymes in cells is controlled by differential gene expression, which in turn conditions the infinite variations of cells, properties, and functions in living bodies.

Among the infinite variety of possible biochemical reactions, enzymes selectively catalyze only well-defined reactions, exclusively transforming specific substances into the right products. This is the main mechanism of the enzymatic administration of all metabolic and physiological processes. Enzymes are a kind of molecular machine that selectively captures certain molecules from millions of possibilities, rapidly and promptly processing them, and then releasing only the necessary mature products. Generally speaking, enzymes are many thousands of times larger than the molecules they transform. If there were no enzymes, the chemical reactions would behave chaotically and at very slow rates which would

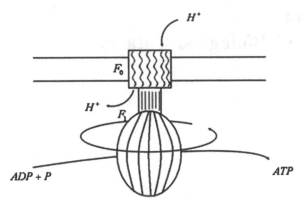

**Fig. 14.1** Enzymes are molecular machines, the main tools and mechanisms of living bodies. This diagram represents the structure of one of the key enzymes, namely ATP-synthase, which continuously synthesizes ATP, something it would be absolutely impossible to do outside of cells. $H^+$-translocating ATP-synthase consists of two parts: a membrane-integrated proton channel ($F_0$), consisting of 13 subunits, and a catalytic subunit ($F_1$), which is embedded in the matrix of a mitochondrion. The "head" of the catalytic part comprises three $\alpha$- and three $\beta$-subunits, which contain three active centers. The energy of moving protons ensures the synthesis of ATP. The catalytic cycle is subdivided into three phases, each of which acts alternately in the three active centers. Firstly, ADP and Pi (inorganic phosphate) are brought to the active center, then a bond is formed between them, and finally the end product, ATP, is released. All three active centers catalyze the next stage of reaction during each transfer of protons through the protein channel $F_0$ from the intramembrane space into the matrix. All components of such extremely complicated nanoscale devices function interactively, incredibly fast, continuously, and with mathematical accuracy

not suffice for the maintenance of life's processes. Furthermore, enzymes ensure a targeted utilization of energy (which is stored as ATP) for specific and necessary cellular processes (e.g., transportation of molecules through membranes, muscle contraction, changing molecular conformations, synthesis of macromolecules, etc.). Some enzymes are energy converters that transform energy from one form to the other. For example, ATP synthases transform the energy of proton transport through the membrane into energy stored within ATP chemical bonds (Fig. 14.1).

Enzymes can be coupled if two or more biochemical reactions are catalyzed in such a way that a total change in the sum of their free energy provides for the progression of the process in a favorable direction from a thermodynamic perspective. In this case, enzymes are linked and accelerate spontaneous and thermodynamically unfavorable processes. This principle of the connection of biochemical reactions is also vital to the basic cellular metabolism and to the overall existence of living bodies.

Depending on the type of catalyzed reaction, the tens of thousands of different enzymes can be divided into six main classes:

1. Oxidoreductases—catalyze redox reactions.
2. Transferases (or transferring enzymes)—accelerate transfer of functional groups of atoms between molecules.

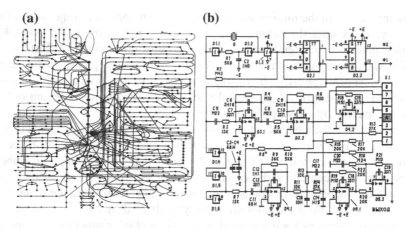

**Fig. 14.2** The basic pathways of cell metabolism form a network analogous to electronics. **a** Classic schemes of structure-functional blocks and metabolic fates of cells. Each point is a specific chemical reaction and its enzyme. The scheme shows pathways representing the transformation of matter and energy as determined by enzymes. The main merit of enzymes is their ability to seek out specific substrates from a large selection and to increase the probability of their almost impossible reaction processes by a large factor. The complex of thousands of enzymes in a cell forms a global structure-functional matrix system of selective catalysis. **b** Scheme of the structure-functional system of an electronic device, represented by a matrix of selective paths of conduction of matter, energy, and information. By analogy with the cellular matrix, it consists of structure-functional blocks connected by communication channels. Each element of the matrix performs its own specific task. The interrelated work of thousands of elements allows for the performance of incredible processes. We may interpret the nanocybernetic colloidal matrix of cellular protoplasm in this manner. It is a standard and extremely complicated system of material-energetic interactions whose expediency and accuracy are ensured by genetic and structural information. Each enzyme can be viewed as a peculiar transistor which transforms, amplifies, and directs flows of matter and energy under the control of intracellular information

3. Hydrolases—catalyze hydrolysis reactions.
4. Lyases—catalyze adjunction of atoms by means of bond breakage.
5. Isomerases—accelerate isomerization reactions.
6. Ligases—catalyze formation of double bonds using ATP energy.

Every class has hundreds of variants, depending on the nature of the substrates used and the chemical bonds involved. Thus, there is a sort of division of labor among the molecular machines. Every cell has thousands of different enzymes that catalyze only their own specific reactions. Various types of cells possess unique enzyme compositions, which condition specific characteristics of their functions. Enzymes are located in different parts of a cell, providing an independent behavior to a wide range of biochemical processes. Organelles contain only their own specific enzymes, thereby establishing distinct functional roles for each compartment.

Enzyme complexes, which catalyze several contiguous reactions of initial substance transformation, form polyenzyme conveyors—metabolic chains, e.g., a complex of glycolysis enzymes which are located in specific places within the cytosol, or the complex of eight enzymes of the Krebs cycle (citric acid cycle), which are located in the mitochondrial matrix. The end product of the work of the first enzyme becomes the substrate for the second, and so on. In this way, biochemical processes are significantly accelerated, substrates do not get lost, time is saved by the delivery of essential molecules, and biochemical processes are strictly directed by specific paths of transformation, avoiding the production of useless products. Thus, the flow of energy and matter is directed by definite and orderly pathways 'paved' by enzyme globules (Fig. 14.2).

Peculiarities in the metabolism of various cells are determined by their different enzymatic compositions. In turn, their protein compositions are determined by differential gene expression. The rate of the general flow of matter in metabolic processes remains constant and is controlled by key enzymes. The slowest enzyme in a chain of biochemical reactions acts as a red light, controlling the quantity of product formation. Such enzymes are usually controlled by cells or an organism. Other groups of enzymes are regulated by the amount of product using a concept of negative feedback—low concentrations stimulate the protein, and high concentrations decelerate it. In this way, an accurate regulation of the rate and quantity of product production is achieved in cells, along with the directionality of chemical processes and flows of matter and energy.

Certain non-proteinaceous compounds may also act as biological catalysts. For example, some RNA molecules possess the ability to catalyze hydrolysis of phosphodiester bonds in nucleic acids. They are called ribozymes and their role in biocatalysis is not yet well understood. In addition, the splicing of some protozoa occurs without participation of proteins, but by RNA. Existence of ribozymes also confirms the theory of evolution, because it shows that initial life forms could have existed on the basis of RNA alone, while proteins and enzymes appeared later on. This stage of development was called the RNA world by Walter Gilbert in 1986.

Many enzymes are compound proteins and consist of a protein globule and a non-protein part (coenzyme). It is the coenzyme that provides a contact between the protein and a substrate during catalysis, 'loosening' its chemical bonds and making it more reactive. Many vitamins and metal atoms are coenzymes. Usually coenzymes are located in the active center of an enzyme—a small area of the protein molecule where fixation of substrates and their transformation into reaction products occurs. Many enzymes also have a regulatory center—an area that provides regulation by other molecules, for example, by reaction products. The rate of biochemical reactions is increased by raising the probability of impact of reacting molecules (substrates) in the active center, and also by accurately orienting the interacting molecules, thereby decreasing the energy barrier. This lowering of the barrier occurs due to: (a) maximal convergence of substrates, (b) the effect of atoms of the active center on definite atoms of the substrate, and (c) changing the energy of electrons in the reacting atoms.

Enzymes are characterized by the following properties: (a) most enzymes are globular proteins; (b) they increase the reaction rate but remain unconsumed themselves, though they do undergo reversible conformational transformations; (c) enzymes catalyze only energetically possible reactions; (d) enzymes possess specificity, that is, a specific enzyme usually catalyzes only one type of reaction;

(e) a minor amount of enzyme transforms a large amount of substrate. One enzyme can transform millions of substrate molecules into the relevant product; (f) most enzymes catalyze chemical processes under mild conditions, e.g., at normal pressure, constant temperature range (0–38 °C), and neutral environmental pH, though some operate more optimally under extreme or stressful conditions; (g) enzyme activity is regulated and depends on the temperature, pressure, environmental pH, substrate concentration, and product concentration; and (h) the rate of an enzymatic reaction is a linear response to the amount of enzyme (at saturation of substrate).

In much of the biochemical and cytological literature, enzymes are primarily only credited with two very important functions—significantly increasing biochemical reactions and regulating metabolism. However, as described above, enzymes also possess a range of more global functions through the basis of catalysis:

a. They ensure increased rates of biochemical reactions;
b. They increase the probability of implementation of only those biochemical reactions that are necessary for a cell among millions of possibilities;
c. They condition the extreme selectivity in the flow of chemical reactions;
d. They provide for purposeful and economical utilization of matter and energy;
e. They take part in transformation of one form of energy into another;
f. They are the key connecting link (transforming link) between the genome and the living body (between the genome and phenome), and also between information and its material incarnation;
g. They ensure directivity and intensity of biochemical processes and, by means of genetic control, regulate metabolic fates;
h. They couple spontaneous processes with the reactions necessary for metabolism.

Inside cells, these enzyme properties effectively create artificial negentropic conditions. Such conditions are characterized by:

1. High probability of realizing unlikely processes;
2. Selective transformation of only a limited amount of universal molecules and only into a limited amount of products;
3. Flows of matter and energy are directed strictly by limited metabolic fates.

All these traits create and maintain order, homeostasis, and the functions of cells and organisms. Thus, enzymes are not only accelerators, but also initiators, sorters, regulators, providers, constructors, and renovators of the molecular world of cells.

It is obvious that enzymes perform work. In particular they grasp the substrates, fix them, change their conformation, transfer molecular parts in the active center, etc. Of course, performing such work requires lots of energy. In this respect, many enzymes use the energy from the chaotic motion of the surrounding molecules, energy of heat motion, and fluctuation of their own molecules for the purposes of performing work. Molecular systems that function on the basis of this concept of work are called Broun's machines. They are micromechanical constructions with parts moving relative to each other under the impact of heat fluctuations. Due to structural peculiarities, only certain parts of these macromolecules possess the ability to move. It is most likely for this reason that a certain selectivity occurs in the fluctuations of enzymatic parts (e.g., only certain segments of the active center), thereby providing the work needed for substrate processing. In this way, enzymes may use the free thermal energy of a system for rapid and directed transformation of the necessary substances.

It should be noted that enzymes use the laws of thermodynamics to perform their work. It is known that catalysts, including enzymes, help spontaneous reactions to progress faster by providing easier paths or by means of mechanisms for lowering the energy barrier. Spontaneous reactions are conducted on the basis of thermodynamic laws and are directed towards increasing entropy, disturbing a given equilibrium, and establishing a new one. Enzymes do not change an equilibrium constant during their work, but they do increase the rate of achieving such an equilibrium under given conditions. This is the main property of catalysts. They merely bring reactions to the same state of equilibrium that they would eventually achieve anyway without catalysts, but in a much faster way. Enzymes allow for the transformation of slow spontaneous reactions into fast ones. In the final reckoning, this results in dissipation, increased entropy, and lower free energy in a system.

Many reactions leading to the general thermodynamic destruction of biosystems are joined together and accelerated by enzymes. However, this dissipative material-energetic flow may also significantly intensify and boost processes that oppose destruction and are directed towards synthesis and creation. In the case of coupled reactions (i.e., simultaneous and interdependent realization of reactions with thermodynamically different directions), coupled enzymes transform the potential energy of spontaneous reactions into the energy required for thermodynamically disadvantageous anabolic processes to occur, thus maintaining cellular order and organization. Coupling is achieved mainly by means of ATP processing (Fig. 17.3). It is in this way that many enzymes prevent the accomplishment of thermodynamic equilibrium, and thereby prevent the destruction of various biological systems.

The specificity of enzymatic work depends on their conformation, which is determined by the amino acid sequence. This in turn is further conditioned by the sequence of nucleotides in the DNA of genes. Thus, the comprehensive process of gene expression includes not only transcription, processing, translation, and modification, but also catalysis. Selective biological catalysis appears, therefore, to be one of the strategic mechanisms for realizing genetic programs in order to establish the phenotype and hence realize all life processes.

There is no single body without any enzymes, and no manifestations of life are possible without them. Enzymes are molecular machines which carry out impossible processes through the mechanism of catalysis, accelerating chemical reactions many thousand-fold. Taking into account the fact that a 10 °C temperature increase promotes a two- to fourfold acceleration of chemical reactions (Vant-Goff rule), we can estimate that, in order to accelerate a reaction 10,000 times, a temperature of 50,000 °C would be required! Now it is easy to understand what enzymes actually do. They create extremely improbable conditions for the ultra-fast behavior of chemical processes at very low temperatures, close to the temperature of water solidification and quite close to absolute zero. That is, living bodies are characterized by unnaturally high rates of selective reactions at ultra-low temperatures.

Enzymes may be considered as the molecular robots of cells, the owners and producers of which are DNA molecules. It is the enzymes that construct the complicated architecture of cells and organisms according to the "blueprints" of genetic programs. Without enzymatic work, no vital activity is possible. If DNA is the legislative basis of life, then enzymes are the executive basis.

## 14.2 Biological Oxidation

The reactions of biological oxidation provide a gradual and discrete extraction of energy from organic substances. The process of energy extraction in animals can be conventionally divided into three different stages. First, at the preparatory stage, macromolecules of food (proteins, polysaccharides, and lipids) are split into monomers (amino acids, monosaccharides, fatty acids) with the participation of digestive enzymes. Then, within cells, this process continues with the participation of intracellular hydrolases. At the second, or anaerobic stage, monomers are partially oxidized, forming several key intermediate low-molecular weight products, mainly through the action of the acetyl coenzyme A and several carboxylic acids. At the third, or aerobic stage, which occurs in the mitochondria, hydrogen is removed from the above-mentioned organic substances by means of special enzymes, and the hydrocarbon skeleton is eventually oxidized into $CO_2$. The segregated hydrogen (universal fuel) then links with oxygen, forming the end product of oxidation—water. The release of energy carriers in oxidation chains is implemented partially and gradually—hydrogen after hydrogen, electron after electron. Finally, energy is accumulated in small ATP molecules, which are a convenient cellular form of energy storage and utilization.

Oxidative processes may be associated with: (a) dehydration–removal of hydrogen from the oxidized substrate; (b) loss of an electron by the substrate; and (c) attachment of oxygen to the substrate. Oxidative processes are always conjugated with reduction reactions, meaning that one substance is oxidized (i.e., gives away an electron) and the other is reduced (i.e., accepts this electron).

The most widespread type of biological oxidation is enzymatic dehydrogenation—the removal of hydrogen. If oxygen is the acceptor, those reactions are called aerobic oxidation. If the acceptor is a substrate of another nature, then such reactions are called anaerobic oxidation. The oxidation of acetyl-CoA in the Krebs cycle is effectively the process of generating protons and electrons, where the acceptors are the coenzymes known as NAD- and FAD-dependent dehydrogenases. During these reactions, electrons with high energetic potentials are transferred by enzymes from reduced NADH and $FADH_2$ coenzymes to oxygen by

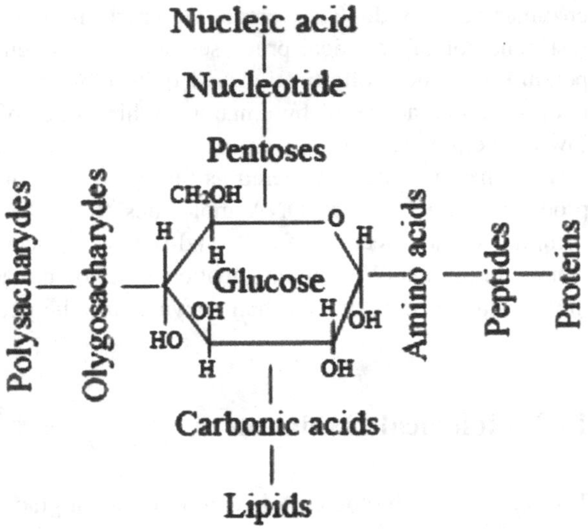

**Fig. 14.3** On the basis of glucose, cells can manufacture virtually the whole range of necessary substances. It should be noted that the amount of possible substances is limited by the quantity of available enzymes. In turn, the qualitative and quantitative composition of enzymes is controlled by the genome

means of electron carriers within the mitochondrial membrane. The reduction of oxygen results from bonding of electrons (which came from the respiratory chain) and hydrogen ions. With the attachment of 2 electrons and 2 protons to oxygen, one molecule of water is formed. The process of oxidation of organic substances in cells, which is accompanied by the consumption of oxygen and formation of ATP and water, is called tissue respiration, and the electron transport systems that carry electrons within the mitochondrial membrane constitute the respiratory chain.

The energy gained by the movement of electrons through the respiratory chain from one carrier to the other is used to transport protons from the mitochondrial matrix to the intermembranous space, where a high concentration is produced. This creates a high electrochemical gradient at the inner membrane. The generated potential is used to synthesize ATP by ATP-synthase, a complicated transmembrane complex, during the process of proton flow through this complex (Fig. 14.1). The generated ATP is used in numerous reactions for the purposes of biological creation and implementation of various functions.

Oxidation–reduction processes are performed by enzymes, which are the executives of genetic programs recorded in DNA. Therefore, biological oxidation is a condition for maintaining the integrity of living bodies, which are the phenotypic frameworks of genomes.

## 14.3  Biological Creation

Cells of living organisms possess the ability to control the formation of various organic molecules, that is, create one substance from others, often of a completely different nature and with completely different properties (Fig. 14.3). Such a property of controlled creation is unique to living systems, or may occur under conditions created artificially by humans. The microscopic volume of a cell contains metabolic conveyors for the production of amino acids, triglycerides, phospholipids, nucleotides, polysaccharides, proteins, nucleic acids, and many other complex molecules. From the trillions of possible variations of organic molecules, a strictly selective formation of only a few thousand occurs. Moreover, metabolic fates are organized in such a way that there is virtually no formation of odd products or by-products. The majority of intermediate substances are used right away at the next stage of a given metabolic conveyor, or become substrates for other metabolic fates. Most of the 'dead-end' organic molecules are also used—they are oxidized to form carbon dioxide and water, which extracts their chemical bond energy.

The majority of organic molecules are formed for the cell's own purposes. However, many cells are capable of producing substances for 'export'. For example, hepatocytes synthesize many proteins for the blood plasma of mammals. These proteins (mainly albumins and globulins) are secreted into the capillary lumen. They are necessary, in particular, for the maintenance of osmotic blood pressure and the creation of optimal conditions for the exchange of substances between the blood and cells of various tissues. Fibroblasts of connective tissue synthesize collagen and elastin, which are secreted in large amounts into the intercellular space, leading to the formation of cartilage, bones, tendons, etc. Likewise, cells of the endocrine tissues produce various hormones. Hormones are biologically active molecules of various natures and structures which cause specific physiological reactions upon interacting with target cells. It is amazing that, of the infinite number of possible variations of chemical substances that possess biological activity, only a few thousand are actually created.

The striking selectivity and specificity of biological creation is ensured by the presence of certain enzymes that catalyze only strictly determined biochemical reactions. Thus, the flow of substances and energy is rigorously directed to a certain 'path', which is paved in an orderly way by protein–enzyme globules. The products of the first enzymatic reaction are the substrates for the second, the products of the second are substrates for the third, and so on. Therefore, there is almost no generation of unnecessary products, since the probability of such formations without catalysis and energy consumption is extremely low. Creation processes in cells are controlled by the genome by means of a selective expression of only the necessary enzymes for each specific process.

The processes of biocreation are energy dependent and interfaced with the hydrolysis of ATP. This couples exothermic biochemical reactions with anabolic reactions, which tend to consume energy.

In plants, the primary synthesis of organic substances from nonorganic ones occurs with the assistance of light energy (transformed into ATP) and the reductive potential of $NADPH_2$ (Fig. 17.1), which serves as the foundation for the existence of millions of species of living organisms:

$$CO_2 + H_2O \xrightarrow[\text{Chlorophyll}]{\text{Light energy}} [CH_2O] + O_2$$

It was the plants that conditioned the emergence of an enormous amount of biomass as a result of the synthesis of organic molecules. This became the basis for the emergence and sustained existence of other organisms that use this biomass for nourishment, growth, and reproduction. The vegetative mass contains an excess amount of organic molecules which are necessary for the survival of many other consumers, from single-celled organisms and fungi to mammals, and, in particular, humans.

Phytophagous organisms feeding on plants obtain organic substances that are perfectly appropriate to their function and organization. There is no need to spend additional energy on the synthesis of organic substances from nonorganic ones. It is sufficient to parse (digest) organic molecules from plants to the smaller and less specific substances from which new proteins, nucleic acids, etc., can be synthesized. Virtually all of Earth's inhabitants depend directly or indirectly on plants as the source of organic substances for nutrition as well as the source of oxygen for respiration. This is why, for example, a collision between our planet and a meteor would lead to a global biological catastrophe, since this would very likely result in solar screening and a prolonged cold night, and this in turn would cause many plants to die, along with many other organisms that cohabit with them. Such an event may well have led to the destruction of most of the dinosaurs and many other less conspicuous creatures. Thus, the primary synthesis of organic substances on the basis of photosynthesis underpins the existence of most living organisms.

Proteins and nucleic acids are the basic macromolecules that are constantly synthesized by cells in massive amounts. To promote the rapidness and accuracy of this process, organisms use the matrix method of reproduction. Nature uses amazingly simple and standard biochemical mechanisms for the synthesis of polymers from a small number of monomers. Macromolecules are then assembled from the small precursor molecules by multiple repetitions of polymerization reactions. Proteins, nucleic acids, polysaccharides, and other macromolecules are synthesized this way. Cells constantly produce large amounts of such necessary molecules.

Naturally, a large amount of overproduced and worn out molecules also undergo controlled disintegration. For example, the lifetime of cellular proteins is only a few hours. After this period, such proteins are destroyed by proteases. Most of the amino acid monomers are used again by enzymes for the synthesis of new proteins, but part of the amino acid pool is further degraded by other enzymes to ammonia, water, and carbon dioxide, which are rejected by the organism.

The presence and work of enzymes, and therefore of all processes, is controlled by the genome. Thus, a rigorously specific set of complementary organic molecules is established and maintained within the organism by means of creative and disintegrative processes which are all controlled by the genome.

# Chapter 15
# Strategy of Copying

## 15.1 Copying and Cloning

Multiple copying of organized structures, systems, and processes is typical for living bodies, particularly for cells. Copying is the process underlying the production of filial creations which are identical to parents. In essence, the process of cloning is the basic process of copying, but continuously repeated. Cloning is based on matrix processes (e.g., replication and transcription) when many identical products are formed on the basis of a single matrix. Cloning results in fast replication and spread of standard units. This concept of 'copying oneself' preserves and propagates, in time and space, the most successful evolutionary achievements. It is used for the purpose of rapid, multiple, and accurate reproduction of the typical structures, systems, and processes, which have a determinative significance for living bodies.

Processes of copying and cloning are inherent to various levels of organization. In particular, there is copying of DNA molecules during the processes of replication (Fig. 20.1), copying of RNA molecules during the processes of transcription, and copying of polypeptides during the processes of translation. In all these cases, molecular complexes are being copied. In cells, for example, the assembly of standard membrane systems, ribosomes, and spindles, and formation of chromosomes, centrioles, cytoskeleton, actino-myosin complex, etc., occurs from one generation to another. In such events, organelles are being copied. Therefore, not only is the genetic material present in each cell during the separation of the daughter cells, but so is a certain part of the ordered inner content, including organelles, highly-organized cytosol, ribosomes, etc. When this occurs, whole cells are being copied. The zygote is the mother cell of all the cells within a multicellular organism. Due to the differential expression of the standard genome, clones of structurally and functionally diverse cells are formed, and their complexes form tissues and organs.

Processes and mechanisms are copied and replicated by copying structures and systems. For example, ordered structures (such as mitochondria, ribosomes,

G. Zhegunov, *The Dual Nature of Life*, The Frontiers Collection,
DOI: 10.1007/978-3-642-30394-4_15, © Springer-Verlag Berlin Heidelberg 2012

endoplasmic reticulum, etc.) get into the daughter cells during cell division. These organelles already contain all the necessary enzymes and other conditions required for the functions of their strictly defined processes.

Because many processes of copying and cloning in biological systems are based upon matrices, where basic maternal structures are used for the formation of thousands of identical elements, we will now expand upon some of the main matrix processes.

*Replication.* The copying of DNA molecules is called replication. During replication, each strand acts as a matrix. DNA replication leads to increasing quantities of the standard genetic material that serves as the base for cloning genomes and cells. This means that the foundation for prolonged existence of separate individuals and populations within all species of living organisms is the matrix copying of DNA and genomes. Every species must maintain the constancy of its own genotype and phenotype. For this purpose, it is important to strictly constrain the invariance of nucleotide sequences within chromosomes. That is why DNA molecules must be duplicated with amazing accuracy before every cell division. The main functional purpose of replication, therefore, is to provide numerous offspring for several generations with stable genetic information for development, functioning, and reproduction.

*Reparation.* In order to preserve the basic characteristics of cells and organisms of a given population, the structure and stability of genetic material must be accurately maintained for thousands and millions years, despite the impact of various mutagenic factors. There are several reasons for the high stability of DNA structure and functions. In fact, it is partly through the durability of the DNA molecule itself, and partly due to the availability of special repair mechanisms for unwarranted modifications. DNA exists as a double coil, and, in the case of accidental damage to one of the chains, replication enzymes are capable of returning that particular segment to a normal condition using the information contained in the undamaged chain. The wide range of different replication enzymes perform constant 'diagnostics' on the DNA, which lead to the removal of damaged or modified nucleotides. Then, by copying the second matrix chain, the DNA structure is repaired, restoring the coded information.

*Transcription.* Proteins are the primary structural, functional, and regulatory molecules, but they become worn out quite fast during vital activity. Therefore proteins must be synthesized again in massive amounts and within a short period of time. The mechanism of matrix copying of protein molecules was created for this purpose. It consists of two stages: synthesis of RNA on the DNA matrix templates (transcription) and synthesis of polypeptides on the matrices of mRNA templates (translation).

Molecules of DNA within each cell contain information for the synthesis of all the necessary proteins. They pass on information concerning the structure of proteins by means of specialized RNA molecules, which are formed by copying from certain segments (genes) of DNA chains. One molecule of RNA after another is copied from this matrix at an intense rate. The transfer of information from DNA to RNA is called transcription. It is a complicated process involving several stages and occurring with the participation of many different enzymes. The process of reverse transcription is known for RNA viruses: synthesis of DNA on an RNA matrix, that is, reverse template copying of information.

*Translation.* The process of transferring information from RNA molecules to the ordered structure of amino acids in polypeptide chains is called translation. It is a process for synthesizing polypeptide chains on ribosomes, following instructions recorded from the genetic code in molecules of mRNA. It is also a matrix copying process, where the informational template is a specific molecule of mRNA, and where numerous copies of polypeptides are synthesized at a very high rate. The copying process goes on until the required level of saturation of a given protein is achieved within a cell.

The copying and replication of proteins is the main mechanism in the realization of genetic programs in particular bodies and processes. Matrix processes allow performing of the extraction of genetic information and its realization in numerous copies, necessary for the cell, with amazing accuracy, high rate, and great efficiency. That is, the phenomenon of copying underlies the rapid construction of conventionally organized living bodies.

It should also be noted that matrix processes prove the link between the genotype and phenotype of the organism: Genotype → copying and cloning processes → phenotype. In other words, a body develops progressively on the basis of these processes. Copying processes also provide the link between present and future genotypes and phenotypes: maternal genotypes → copying processes → daughter genotypes, or maternal phenotypes → copying and cloning processes → daughter phenotypes. Thus, the offspring emerge as a result of copying and further cloning of hereditary information.

Increases in entropy variability are spontaneous and imminent. In order to be stable and not disappear, living organisms must fight constantly against increasing entropy, and this is only possible for a certain period of time. Therefore, for a strong and prolonged existence, it is necessary to reproduce copies of oneself from time to time, forming either exact copies of cells which arise from cell division, or non-exact copies which arise through organismal reproduction. Copying and cloning is the foundation for preserving and disseminating material organization and information, and is typical only of living bodies. Nevertheless, a direct participant in and administrator of all the listed events is the genome, which constantly pursues its own selfish interests, and these in turn affect the whole organism.

## 15.2 Natural Selection

The diversity of organisms, their properties and characteristics, and their expediency of organization and functioning are conditioned by natural selection of the best adapted living bodies and their genomes. Natural selection is a constantly running process which, in its external manifestation, is expressed through the maintenance of organisms with useful properties for the given conditions, and through the elimination of representatives that are less well adapted to the environment. Natural selection has always been and still is one of the main mechanisms of evolution, along with genetic variation and the struggle for existence.

Ecological niches are inhabited by a wide range of phenotypically different organisms possessing very different genomes. The fitness of phenotype representatives (and, therefore, genotype representatives) is varied. Selection of the fittest organisms proceeds along the path of selection and fixation in the population of those phenotypic characters that increase the chances for survival. Since all characters are determined by certain genes, alleles, and their combinations, the genomes of the fittest organisms are thereby selected via the natural selection of phenotypes. That is, natural selection has a molecular-genetic basis. Its result is not only the survival of the fittest organisms, but also the enrichment of the gene pool of the population with beneficial alleles.

The virtually unlimited material for selection emerges as a result of constant reproduction, based on the phenomenon of copying and cloning. The resulting individuals possess diverse genotypes and phenotypes through the mechanisms of genetic variation: mutations, combinatorial variability, hybridization, and transgenesis. Numerous individuals with various combinations of new alleles and characteristics emerge in populations as a result of these processes.

Natural selection occurs at all stages of an organism's ontogenesis. For example, at the pre-embryonic stage of development, during the process of insemination and the early stages of fertilization, only the most valuable and active spermatozoa are selected by environmental conditions from the millions available. At the stage of embryonic development, the dominating mechanism of selection is selective mortality. During this process teratoid, abnormal embryos, which carry modified genomes such as gene, chromosomal, or genomic mutations, are deleted from populations. Many weak organisms die right after birth. Those organisms die selectively that have structural or functional imperfections, usually determined by genomic defects. At the period of reproduction, those organisms that have survived but are defective have much less chance of securing posterity, and thereby passing on their defective genes. As a result, natural selection provides for selective reproduction of genomes. Consequently, favorable characters, and hence also their genes, are accumulated in a sequence of generations, gradually changing the genetic composition of the population in a biologically expedient direction. That is, in Nature, natural selection occurs exclusively via the phenotype, while selection of genomes (which is fundamental) occurs in a secondary manner through selection of the best adapted phenotypes.

As an evolutionary mechanism, natural selection acts within populations. In this respect, individuals are the subjects of action, and specific characters are points of application of selection. The gene pool of a population is changed by changing concrete genomes of particular individuals and increasing (or decreasing) their numbers. This is the main route to the emergence of a new species.

Selection may happen under the impact of any factor that changes genotypes in such a way that, in the struggle for existence, it increases the chances of reproduction of particular individuals that have definite genes and their combinations. Therefore, one real effect of natural selection is to increase the frequencies of those genes in the population that benefit for achievement of reproduction by the organisms carrying them.

In reality all living bodies are amazingly complete and perfect creations. Regardless of the complexity of structure, they are all expediently arranged and well adapted to the inhabited environment. Peculiarities of the genome only affect sizes and shapes, but not the ability for adaptation, survival, and reproduction. In this respect, organisms necessarily possess the required minimum of essential functions for survival. Therefore, every existing genome is perfect in relation to a determined phenotype that fits ideally into its environment.

Thus, as one of the main instruments of evolution, natural selection has conditioned the emergence of many different species of living organisms (genomes and their phenomes) that are very well adapted to their environment. That is, natural selection is a natural mechanism for selection of genomes and recorded information.

# Chapter 16
# Strategy of Self-Organization

## 16.1 Synergetics in Biology

Biological objects are multiheterogeneous, but at the same time complex and ordered systems. Synergetics is a science that studies processes of self-organization in open dynamic systems. It is based on physical and mathematical methods and seeks to define and generalize laws of emergence and development of organized structures. This approach allows scientists to focus on conditions of lability and mechanisms of emergence and rearrangement of structures.

Fluctuation and bifurcation are the main concepts of synergetics. Fluctuation can be treated as an oscillation of the system itself, or an oscillation of the system's elements, around some average value. Bifurcation is a certain critical threshold point of fluctuation when a system is in some sense in two modes at the same time. If a system reaches a bifurcation point, it may cause a qualitative change in its condition and behavior. The fluctuation may spasmodically increase at this range, whereupon the subsequent behaviour of the system will become indefinite, in the sense that it becomes impossible to predict whether the system will return to its initial mode or a qualitatively new mode will appear. For various reasons, new organized structures may appear in chaotic, disordered, non-equilibrium systems as a result of fluctuations and bifurcations. An ordered structure is an object, system, or part of a system that possesses resistance and rigid binding, and hence has the ability to resist external or internal perturbations. This is exemplified by the regular crystal lattice of atoms in rigid bodies, or the irregular but highly-organized structure of living organisms, which consist of many ordered elements. As opposed to order, chaos is characterized by an inner homogeneity and by an absence of regularly sited stable structures and their associations.

One example of self-organization in cells is provided by the self-assembly processes of the phospholipid membrane, actinic filaments, and ribosomes. The simplest self-organizing system of biomolecules is an interacting system of an enzyme and its substrate. At sufficient concentrations, self-organization occurs, with transformation of particular organic substances (substrates) into specific

G. Zhegunov, *The Dual Nature of Life*, The Frontiers Collection,
DOI: 10.1007/978-3-642-30394-4_16, © Springer-Verlag Berlin Heidelberg 2012

products. Such self-organizing transformations are very important for the existence of cells.

An example of a self-organizing process on the level of organs is the spontaneous automatism of myocardium contraction. Rhythmic heart contraction is driven by internal rather than external causes. It is due to the rhythmic spontaneous formation of electric potential on the membranes of driver cells which distribute the rhythm to the whole heart. The electric potential is self-organized as a result of the relocation of anions and cations (fluctuations) through the membranes of these driver cells. When a critical concentration of molecules is reached on either side of the membrane (bifurcation), this determines electrical breakdown of the membrane and generation of an electrical impulse.

Along with the processes of self-organization of structures in the dynamics of various open systems, degradation processes also come into play. Thus, systems may be organized in general, but at the same time are in non-equilibrium, and deteriorating. Such systems are said to be dissipative. They can exist for a certain period of time, but only by exploiting free energy from the environment. Cells and multicellular organisms belong to precisely this type of system.

Dissipative structures emerge far from the equilibrium of a system and allow the possibility of transition to an 'organized chaos'. In such structures, unpredictable (that is, random) but organized processes and structures can emerge. Such chaos is described as dynamical or determinate. Determinacy (that is, predestination) manifests itself in the necessary perturbation of a system, and chaos manifests itself in the unpredictability of the places and times at which these self-organization points will emerge. The dynamics of chaos can be interpreted as fluctuations of particles or objects under conditions of chaotic motion (Fig. 16.1). Thus, 3.5 billion years ago, determinate and dynamical chaos in the great oceans could have conditioned the emergence of organized processes and structures of diverse composition, size, properties, and duration in numerous locations of its enormous volume. That is, many bifurcation states must have emerged as a result of fluctuations, causing subsystems to turn into qualitatively new states. The sieve provided by natural selection would have preserved the most stable systems, in which the forces of internal bonds proved to be stronger than external forces. Then, throughout hundreds of millions years, the evolution of these islands of order would have gone on in the ocean of chaos, and in this way protobionts could have emerged.

Self-organizing systems are nonlinear and unpredictable. It is virtually impossible to predict the qualitative or quantitative parameters of an event, or the place and probability of its realization. That is, events that develop in chaotic systems are of an accidental (low probability) nature. A cell, in which the processes of thermal destruction are going on all the time, can be interpreted as a dissipative, but constantly self-organizing system with regard to the processes of anabolism. It is impossible to foresee which chemical reactions (out of billions of possibilities) will occur, nor indeed where they will occur. However, the general direction of metabolism is conditioned by the existence of organized structures and enzymes, which selectively catalyze only the necessary biochemical reactions. In

**Fig. 16.1** The worldwide flow of chaotic destruction of the initial order of the material world creates spontaneous sites of self-organization

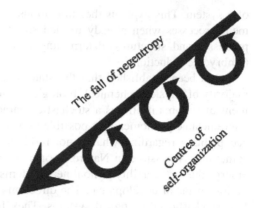

this way the probability of realization of particular processes necessary for a cell is significantly increased. The improbable becomes probable, the unstable becomes stable.

The influence of various factors on the organization in dissipative systems has been established. Factors that have an impact on developing dynamical systems are called attractors. In Nature, these may be the various forces that arise as a result of natural processes: heat, cold, wind, lightning, various oscillations, radiation, etc. In biology, they may be any external or internal factors that have an impact on macromolecules (DNA, RNA, proteins), membranes and organelles, cells, or organisms. Such factors can be physical, chemical, biological, or informational. Taking into account the enormous number, force, combinations of influences, and points of application of these factors, it is clear that the evolution of biosystems could be directed in an infinite number of possible ways. This is exactly what happened in the early stages of biological evolution. And in accord with the theory of I. Prigozhin, the dissipative organized structures not only maintain themselves in the stable non-equilibrium state, but can even develop under conditions of increasing material and energy flows. The mechanisms of natural selection fixed the most stable and thermodynamically efficient organic systems. Their further evolution conditioned the emergence of living bodies, which possess the key characteristics of dissipative systems: the ability to self-organize and develop and sensitivity to minor changes in the environment.

High levels of order in living organisms are maintained by means of evolutionarily developed mechanisms of protection from the impact of unwanted internal and external factors. In particular, the processes of self-repair, selective catalysis, and selective degradation are ensured by constant consumption of free energy and its transformation. On the other hand, a small set of attractors of chemical nature is used by organisms to have deep and purposeful impacts on various biological systems (e.g., dozens of hormones, some neurotransmitters, ATP, cyclic adenosine monophosphate, acetyl coenzyme A, and some other impact factors). Attractors could also be those structures that become more favored than others under the processes of development, self-organization, and evolution of a system. This explains the stipulations and directivity of embryonic development processes when already formed structures have an impact on surrounding processes and structures, determining their development in a certain direction (embryonic induction).

It has been established that the state of chaos in evolution is typical for the majority of physical, chemical, biological, and social structures, and the development of systems itself has a stochastic character. That is, any type of absolutely unpredictable phenomenon is possible in Nature. There is a sort of presumption of acceptability regarding what is not forbidden by the laws of natural science. If something is possible in Nature, sooner or later it will happen. Such states can emerge due to the ability of nonlinear systems to evolve in diverse ways, choosing various paths of development. The directions of evolutionary processes are therefore stochastic in biological systems. They have developed chaotically and randomly over billions of years. The directions of evolution and the emergence of living organisms were also determined by random actions of certain attractors and environmental conditions, and later by nucleic acids—first RNA, then DNA. DNA became and still remains a superattractor, capable of having a dramatic impact on structuring the surrounding chaotic material. But the molecule of DNA also emerged randomly and is prone to the random impact of various factors, undergoing random mutations, which are in turn selected by environmental conditions. Thus, millions of species of living organisms that inhabit the Earth have random genotypes and phenotypes. Moreover, they evolve in a random and unpredictable manner now, and will continue to evolve in this way later. In the event of a possible repetition of the evolution of certain living organisms, the process will behave in a completely different manner, following an absolutely unpredictable path.

This assumes that life is not only a phenomenon of the existence of certain autonomous organisms and species, but a continuous process of evolution of the Integrated Life System, comprising an unending network of genomes and phenomes. Molecules of DNA and their combinations, depending on accumulated genetic information, are able to direct the processes of organization to a certain point, forming around themselves a structured material space which ensures their own survival and reproduction. This structured space eventually gains the appearance of a certain organism—a phenome forms around the genome. The phenotypic manifestation of life as individual living bodies is what we see around us and perceive as the true manifestation of life. But within the framework of billions years of evolution, they are just temporary forms of existence of constantly evolving genotypic life. The direction and goal of this development is ambiguous, random, and unpredictable. It is just one form of the existence and development of matter.

Model systems, by means of which scientists describe the surrounding world, contain elements of order, as well as disorder. From this standpoint, the model of dynamic chaos is a link between completely determined systems and fundamentally random ones. On this assumption, it is obvious that chaos at the micro level may lead to order on the macro level (e.g., chaotic molecular motions in cellular cytoplasm provide the foundation for the ordered metabolism of cells). That is, in many biological systems, order and chaos go hand in hand, and chaos itself acts as an ultra complex form of order. Moreover, increasing order lowers the possibility of development of a system. For example, less differentiated (less ordered) embryonic stem cells can develop into dozens of variations of other cell types. In contrast, highly ordered differentiated cells (neurons, myocytes, etc.) are virtually incapable of reproduction and development. Thus, chaos, with its dynamic instabilities, provides the potential driving force for self-organization of a system throughout its development process.

So what defines the single form of self-organization actually adopted among the many possible directions? It is known that, in non-equilibrium transitions, that is, at bifurcation points, where the processes of self-organization occur, a system's behavior corresponds to the one with the least production of entropy. That is, during the process of evolution of living systems, the direction of their development will be defined by the appearance of structures with maximal order under given conditions. It can be said that biosystems structure matter and energy from the environment, so that an ordered part remains in the system, and a disordered part returns to Nature. As one of the main characteristics of life, order has emerged from the chaos of the material world through the process of self-organization, which is based on the physical and chemical laws of interaction of molecules and their systems.

The property of self-organization is inherent to a system independently of the physical nature and peculiarities of that system's structure. Both organic and inorganic ordered equilibrium as well as non-equilibrium systems can be formed. The ability to self-organize is inherent not only in the structures, but also in the processes. In particular there is a theory of hypercycles that allows the spontaneous formation of self-maintaining non-equilibrium networks of enzymatic reactions, their development by means of a feedback system, and the generation of new forms of organization.

So one of the main properties of the developing material world is its ability to self-organize on the basis of the laws of synergetics. The foundations of synergy allow the realization of new patterns of organization in developing systems, constructing complex structures from simple ones, and integrating the whole from its parts. At the same time, the whole is never equal to the aggregate of the parts, but appears to be qualitatively different. Patterns of self-organization allow a new understanding of processes in Nature, affecting the integrity of the entire material world and treating life as a manifestation of evolving matter.

## 16.2 Determinate Self-Organization

Living bodies are distinguished by highly ordered structures and processes, organized into a single system. Underlying the appearance of order in biosystems are processes of self-organization based on the principles of synergetics. Self-organization is the process of spontaneous ordering of elements of a system under the influence of various environmental factors. For example, phospholipids in an aqueous environment self-organize into micelles or vesicles as a result of their amphipolar properties. Moreover, order and organization can arise on the basis of information about the content and order of the arrangement of components of a system. In other words, information provides an additional determination of and guidance for the processes of self-organization in a definite direction. Thus, living systems are capable of determined self-organization—the formation of standard ordered structures and processes based on genetic information and the mechanisms of synergetics.

Genetic information is recorded in molecules of nucleic acids. It is realized in the process of individual development. As a result, order develops in the highly organized structures of living bodies on the basis of "genetic blueprints". However, DNA genes only contain information about the primary structure of proteins and the molecular intermediaries of their synthesis—RNA. An enormous amount of further information is lacking in the genes themselves. In particular, it is still not known how this information is recorded and where it is located. Where is the information about the spatial structure and functions of proteins? How is information recorded regarding the forms and sizes of cellular structures? Where is the information about the pathway by which biomembranes should assemble themselves from concrete lipids and proteins, and about the forms and sizes of cells? Why do certain cells combine into tissues and organs of a definite form and size? What information determines the location of an organ or body parts? And many other issues remain. Very likely, the additional information needed for self-organization arises in biological systems during the process of development. For example, the formation of spatial structures of proteins from linear polypeptides is determined by the qualitative and quantitative composition of amino acids in the process of synthesis. The synthesis and folding of contractive proteins in muscle cells predestines them for the formation of contractive tissues. In other words, developing dynamic living systems are determined self-organizing systems. They are characterized by low values of entropy, i.e., they are in a state of thermodynamic non-equilibrium. This unstable state is maintained through two competing processes. On the one hand, in biological systems, the process of thermal destruction of order occurs constantly, and on the other hand, this order is immediately restored genetically by controlled flows of energy and material passing through these systems. Ordering processes in self-organized systems are accompanied by the utilization and dissipation of energy.

Biological systems are characterized by the thermodynamically stable state of their elements and a steadfast resistance to aggressive environmental factors. This is possible if the energy binding elements of the system exceeds the energy of external forces acting on the system from the environment. Organisms prevent their destruction at the expense of a constant influx of energy and matter from outside, and this is an essential condition for the existence of non-equilibrium systems. If the ability of a system to self-organize and self-support its structure weakens, for example, due to inadequate energy replenishment, then its elements become less organized and the system gradually falls apart (its entropy increases).

Living bodies actually act as highly organized systems in reducing their own entropy at the expense of an increase in the entropy of the surrounding environment. It may also be supposed that the excess free energy absorbed by an open system can promote its self-complication. In other words, living organisms in the process of development not only resist the increase of entropy and chaos inside themselves, but also form more and more complex structures by utilizing energy, material, and information from the external environment. The complication and perfection of biological systems occur precisely on this basis.

Interaction and association are fundamental mechanisms of self-organization at all levels. For example, aggregation at the molecular level unites separate units into groups and forms specific complexes of macromolecules (e.g., enzymatic complexes). It is known that amphipolar molecules of phospholipids in aqueous environments form ordered structures that are prototypes of biomembranes. The association of various organelles and their coordinated functioning provides for the cellular level of organization of life. In turn, cells that vary in terms of structure and function can successfully combine, interact with each other, and form functional units of tissues and organs.

The most impressive example of determined self-organization is the process of development of embryos, in which—over a relatively short period of time—one cell (zygote) transforms into a large multicellular organism, billions of times bigger, comprising trillions of cells and consisting of hundreds of cell types that form diverse tissues, organs, and body parts (Fig. 4.2). Embryogenesis is based on the differential expression of genetic information and use of a significant amount of free energy and matter to create order. It is enabled by a complicated process of determination, division, and differentiation of cells under the action of regulatory molecules. Morphogenesis underlies the processes of interaction and migration of cells and their selective division, growth, and death, regulated and directed by special regulatory and informational molecules. Interaction of parts of the developing embryo, based on the interaction of its cellular molecular receptors and signaling molecules, assures the formation of a body with all its morphological and physiological features. In other words, a significant amount of further determining biological information is already exhibited and realized in the structures of the organism in the process of its gradual formation and development. Furthermore, the genotype of the zygote obviously encodes not only information about the primary structure of proteins and intermediaries of their synthesis, but also a colossal amount of information representing "instructions" about the order, directionality, and intensity of their synthesis. In other words, it is not the phenotype itself that is encoded in the genotype, but rather a series of "instructions" regarding its creation and composition based on enzymes that make for self-assembly under certain conditions on the basis of molecules, organelles, cells, tissues, and organs.

Thus, determined self-organization is a process of formation of ordered structures of living bodies based on biological information, free energy, matter, and patterns of synergetics.

# Chapter 17
# Strategy of Matter and Energy Transformation

## 17.1 Exchange of Matter and Energy

One of the main distinctions between living organisms and non-living bodies is a multilevel organization of matter that persists for a long period, despite the requirements of the second law of thermodynamics. This property of living systems is supported by fulfilling three conditions. The first is a constant inflow from the environment of the molecules needed for the synthesis and compensation of worn out, damaged, oxidized, and exploited macromolecules. The second is a constant inflow from the environment of the energy needed for synthesis and maintenance of order among molecules, their complexes, organelles, and cells. The third requirement is that waste products and unused or toxic substances should be rendered harmless and/or discarded from the system into the environment. The above-mentioned processes are very important for living bodies. A complex of these processes is called energy and matter exchange. These are interconnected processes because chemical energy is transferred along with the flow of organic substances. Energy and matter exchange may be considered as a form of interaction of an organism with the environment. The totality of substances entering an organism, along with their transformation and excretion of remains, are called an external exchange, while the totality of internal processes belongs to metabolism.

The required substances and energy enter the organism in the form of food. Proteins, fats, polysaccharides, nucleic acids, etc., are obtained through the process of nourishment. These substances are broken down by enzymes into monomers and the macromolecules necessary for the cell are synthesized from these monomers. Some of the substances are oxidized to $CO_2$ and $H_2O$ and the energy of their chemical bonds is transformed into electrochemical gradients or macroergic bonds of ATP molecules. 'Waste products' such as carbon dioxide, water, and urine are excreted into the environment.

Amino acids, obtained by protein breakdown, are the main construction material for all the proteins of cellular structures and enzymes. Monosaccharides and lipids also participate in forming cellular structures, but their main role is

G. Zhegunov, *The Dual Nature of Life*, The Frontiers Collection,
DOI: 10.1007/978-3-642-30394-4_17, © Springer-Verlag Berlin Heidelberg 2012

energetic. They introduce the energy, 'stored' in the chemical bonds between the atoms of these molecules, into the cell.

Plants consume the energy of light and use it to synthesize primary organic substances by means of photosynthesis, using water and carbon dioxide from the environment. A minor product of this process is oxygen, which is discharged into the atmosphere. For production of energy during the dark period of the day, plant organisms use environmental oxygen for oxidation of organic substances in mitochondria and formation of ATP. In this case the waste products are carbon dioxide and water.

Fungi, single-celled organisms, and bacteria live in a similar way, by means of matter exchange with the environment. For the cells of multicellular organisms an outer medium is the intercellular fluid, from which they obtain oxygen and nutrients, and into which they excrete metabolic waste products. In turn, the intercellular fluid maintains the consistency of its inner medium by means of matter exchange with blood, and the blood system is interconnected with the consumption or excretion organs of an organism.

Thus, energy and matter exchange maintain the consistency of molecular composition, organization, and energetic potential of cells, providing for metabolism and various functions of cells in all organisms. This in turn conditions homeostasis and long-term maintenance of the living body's integrity.

The processes of metabolism are conditioned in the first place by the functions of various proteins. The presence of specialized proteins and their activity depends on selective gene expression. That is, metabolism, as an element of the phenotype, comes under the control of the cellular genome. In addition, protein exchange processes may be regulated by various hormones and neuromediators, another type of indirect genetic control.

## 17.2  Metabolism

One of the main conditions of organismal life is the continual selective chemical transformation of molecules from one substance to another. Hundreds of thousands of different biochemical reactions are simultaneously implemented in cells in a strictly coordinated way. Many biochemical processes are tightly interfaced with biophysical processes. Metabolism is a complex of all the interrelated highly-ordered processes and mechanisms of transformation of matter and energy in a cell.

Organic molecules that enter the cell undergo complex chemical transformations. Every second, thousands of different substances are disintegrated and thousands of others are formed as a result of purposeful enzymatic degradation and the subsequent synthesis of the required macromolecules. The energy of chemical bonds in organic molecules is transformed by means of a series of complex steps into the energy of ATP bonds. High-energy ATP bonds are easily split by special enzymes. This energy can then be used for forming the necessary molecules and performing the many different kinds of work of a cell or an organism. A living organism is an isothermal chemodynamic engine with very high efficiency—up to 60–70 %, using energy from the chemical bonds of disintegrated organic substances at constant temperature.

The complex of processes leading to the enzymatic synthesis of molecules necessary for the cell and the formation of cellular structures from these molecules is called *anabolism*. This includes photosynthesis, the synthesis of proteins, nucleic acids, and phospholipids, and the formation of membranes, ribosomes, etc. These processes are endergonic, occurring only through the use of energy. The complex of selective processes of disintegration of organic molecules is called *catabolism*. In particular, this includes the processes of glucose oxidation during glycolysis, oxidation of fatty acids, deaminization of amino acids, disintegration of worn organelles, etc. The processes of catabolism are exergonic, i.e., accompanied by energy dispersion. In cellular metabolism, the processes of anabolism and catabolism are often conjugated. The main conjugating molecule is ATP. Its hydrolysis occurs with energy release, which is immediately used for endergonic reactions (e.g., at certain stages of protein synthesis).

Interrelated chains of chemical reactions constitute metabolic fates. For example, the Krebs cycle (citric acid cycle) is a complex of eight interrelated biochemical reactions leading to the disintegration of the hydrocarbon skeleton of almost any organic substance into carbon dioxide and hydrogen ions. The enzymes implementing this cycle are compactly located in the mitochondrial matrix. The sequences of the majority of metabolic fates and cycles and the set of participating enzymes is astonishingly similar for all living organisms from bacteria to humans.

One of the features of metabolism is a contiguous recycling of biologically important macromolecules. Recycling is the repeated multiple usage of monomers (amino acids, monosaccharide, nucleotides) that are formed after the controlled disintegration of over-age macromolecules and used to build biological polymers (proteins, starch, nucleic acids). In this way, significant energy saving is achieved, along with the constant renewal of cellular structures.

The entire complex of different biochemical reactions is accurately regulated and coordinated in time and space. The rate, directivity, activation, and disabling of chemical reactions are controlled by enzymes. In general, the metabolism is regulated by controlling the qualitative and quantitative composition of enzymes and their activity. This regulation is achieved by selective synthesis of the required molecules, while the selectivity of a given synthesis is in turn controlled by the genome. Enzymatic activity is altered by reversible inhibition or activation by substrates, products, or hormones.

Biochemical transformations are tightly interconnected with biophysical processes. In particular, the heat motion of molecules is very important for the behavior of chemical reactions. It ensures contact and interaction of molecules and further transformations. The transport and diffusion of molecules into the cell, within the cell, and out of the cell are also achieved by Brownian motion. On this basis electrochemical gradients and potentials are created and energy is transformed. Osmosis is also of great importance for maintaining cellular and tissue homeostasis. Many molecules and supramolecular structures of cells possess physical properties of polarity and hydrophobicity. Cell membranes possess electric potentials and outgrowths of neural cells conduct electrical currents. Energy transformations in living organisms obey the laws of thermodynamics. Biophysical and biochemical processes also condition the functions of sight, hearing, motion, conduction of neural impulses, penetrability of various substances, and many more. So the maintenance of the structure and functioning of cells, tissues, organs, and the whole organism is ensured by interconnected, purposeful biochemical and biophysical processes.

Thus, the intracellular melting pot of metabolism ensures all properties and functions that underlie the lives of individual cells and multicellular organisms. The task of metabolism is primarily to provide a constant interaction of cells with the environment in order to maintain homeostasis in the phenotypic realization of the genome. In turn, from the point of view of metabolism, the phenomenon of life may be interpreted as a system of genetically operated molecular processes, which are strictly ordered and organized in a certain body, and directed to maintain its integrity and interaction with the environment.

## 17.3 Transformation and Utilization of Energy

All living organisms are open, highly-ordered, nonequilibrium systems. Energy is required for the creation, maintenance, and functioning of such systems. Energy is a general quantitative measure of motion and interaction of all types of matter, or a potential to perform work. It reflects quantitative modifications in the condition of bodies, their motion, or structural changes under various types of interactions. The concept of energy brings together all the phenomena of Nature. No physical phenomenon or chemical reaction can be accomplished without energy costs in one form or another. Well known forms of energy are thermal, light, electrical, mechanical, and chemical. The different forms of energy may transform into each other during physicochemical processes, but in all cases, the total energy remains unchanged.

The most convenient form of energy for living organisms is chemical energy, since it is easy to store, transport, and transform from one form to another, whenever required. Chemical energy is the energy of chemical bonds, which arises through interacting electrons. Practically all aspects of a body's life depend on energy transformation among electrons. Such changes are based on quantum mechanisms, which underlie the transformation of matter and energy, and comprise the only source of energy in living systems. Electrons are tiny discrete units of quantized energy and matter. Quantized energy is very convenient for extraction in small amounts from organic substances by removing electrons or protons during oxidation.

Energy enters the cells of animals from the outside in the form of nutrients (mainly carbohydrates and fats). It is stored in chemical bonds between atoms in these molecules. Breaking these bonds leads to release of energy (electrons and protons are redistributed), and this energy is transformed and stored in several forms. (1) Proton potential ($\Delta\mu H^+$) on the inner membrane of mitochondria, chloroplasts, or mesosomes of bacteria. Such a potential is achieved by active accumulation of protons on only one side of the membrane. This form of potential energy of protons can be used directly to carry out certain tasks, such as rotation of filaments or fluctuation of the cilia of single-celled organisms, but it is mainly used for transforming $\Delta\mu H^+$ into ATP. (2) Sodium potential ($\Delta\mu Na^+$) on plasma membranes of cells of single-celled and multicellular organisms. The potential energy of $Na^+$ may be used directly for performing a specific task. For example, the energy of this potential is used to transfer various molecules required by the cell, in particular amino acids, monosaccharides, ions, etc. Significant amounts of energy are stored in multicellular organisms in the form of the membrane potential, because each of trillions of cells possesses potential energy on plasma membranes, as well as on inner membranes. In this way, cells solve many transport problems regarding inflow and outflow of substances. (3) Macroergic bonds of ATP. This is the main way of storing and using energy. In this mode, energy can be used by cells and organisms for performing all sorts of work, including synthesis, transport, motion, etc. In the first place, energy is needed to maintain homeostasis in living bodies. Indeed, this uses up to 90 % of the energy, while less than 10 % is used for the various physiological processes. The energy generated in cells may be transformed from one form to another. For example, the energy of the proton gradient on the inner membranes of mitochondria or chloroplasts is transformed by means of ATP-synthases into the chemical energy of macroergic ATP bonds. Molecular complexes of ATP-synthases use the energy of motion of protons in the formation of ATP (Fig. 14.1).

The emergence of the enzymatic mechanism for mass transformation of energy and mass synthesis of ATP had a whole series of major advantages for sustaining life in living bodies. It led to significant progress in the morphological and functional facilities of organisms, as well as improving reproduction and dispersion, and constituted a considerable evolutionary step.

Chemical energy stored as phosphate bonds in ATP is easily released with the help of enzymes and used to perform work (Fig. 17.3), e.g., mechanical work during muscle contraction, electrical work during conduction of neural impulses, molecular transport through cell membranes, and energy supply for chemical transformations during synthesis of various substances or for cell growth and division.

The mechanism of selective oxidation of certain organic substances with the help of special enzymes underlies energy production processes in all cells. Enzymatic oxidation is a process of forced separation of electrons or protons from various organic molecules. Energy is transferred along with these particles to the composition of other substances, in particular to the structure of NADH and FADH, and then into the bonds of ATP molecules. These molecules are the universal batteries for all living organisms. Energy is thus extracted and stored in a discrete form. This is extremely convenient since it allows organisms to

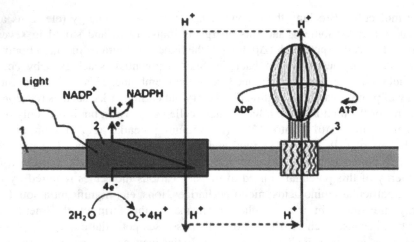

**Fig. 17.1** Transformation of light energy into a cellular form of energy in plant chloroplasts. Absorbed light energy is used to extract electrons and protons from water. The energy of the moving electrons is then used to transport protons into the intermembrane space and create a high potential. The energy of this current subsequently conditions ATP synthesis by means of ATP-synthase. *1* inner membrane of thylakoids, *2* system for transport of electrons and protons, *3* ATP-synthase

progressively accumulate large energy resources and then promptly, and in suitable doses, use it for various cell activities, even in the most inaccessible areas. So cells and organisms are biochemical machines that work at constant temperature and pressure on the basis of chemical energy.

*Basic energy transformation processes in living systems.* The main source of energy for the majority of living organisms is the Sun. The radiant energy of the Sun (photon flux) is consumed by plant chlorophyll and, through a series of complicated enzymatic processes, is transformed into a proton gradient in chloroplasts, and then by means of ATP-synthases into the energy of chemical bonds in ATP (Fig. 17.1).

Accumulated energy is used for the synthesis of primary organic substances from nonorganic molecules of carbon dioxide and water. The chemical bonds of such organic substances finally accumulate the energy of solar photons. The totality of these processes is called photosynthesis. The resulting primary organic substances are subsequently eaten by phytivorous animals and transformed into secondary organic substances in animals. Animals do not possess the ability to photosynthesize and cannot use solar energy directly for the synthesis of the necessary organic substances. For them, the acquisition of organic nutrition, and hence the necessary substances and energy, depends on plants. Herbivorous animals depend directly on plants, while carnivorous animals depend indirectly on plants. Animals also consume oxygen given off by plants. In this way, the energy of sunlight travels from the chemical bonds of organic substances in plants to the chemical bonds of organic substances in animals.

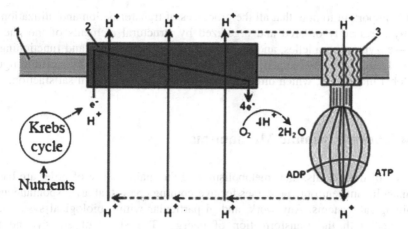

**Fig. 17.2** Mitochondrial oxidation of nutrients into the cellular form of energy. This is an extraordinary nanotechnological mechanism of conjugated oxidation processes with transformation of energy and accumulation of energy in ATP. Processes of oxidation of organic substances provide a source of electrons and protons. The energy of moving electrons is then used to transport protons and create a high potential. The energy of this current subsequently conditions ATP synthesis by means of ATP-synthase. *1* inner membrane of mitochondrion, *2* electron and proton transport system, *3* ATP-synthase. There is a clear similarity in the way energy transformation mechanisms are organized in the cells of plants and animals (Fig. 17.1). The difference lies only in the source of protons and electrons

Organic substances obtained by animals are oxidized in cells and the energy of the chemical bonds in these substances is partially transformed through a series of enzymatic molecular processes, mainly into the energy of phosphate bonds in ATP. The whole complex of these processes in animal cells is called cell respiration. This process occurs in the mitochondria in all cells of every organism (except anaerobic organisms). Such structures, e.g., mitochondria and chloroplasts, which are able to convert one form of energy into another, are called energy converters. They correspond to the generators in a power plant, where the energy of falling water or thermal energy is transformed into electrical energy. Chloroplasts transform the energy of electromagnetic radiation (from the Sun or artificial light sources) into the energy of an electrochemical proton potential on biomembranes, which is used for conversion and storage of energy in the chemical bonds of ATP (Fig. 17.1). Mitochondria convert the chemical bond energy of various organic substances into an electrochemical proton potential on their inner membranes. The energy of this potential is used to cause migration of protons through ATP-synthases and transform kinetic energy into the energy of chemical bonds in ATP (Fig. 17.2).

Therefore, in order to maintain life, organisms must constantly obtain, convert, and use energy. The inflow of energy allows them to function and to maintain a high level of organization for a long period of time. If the inflow of energy to the organism significantly lowers or comes to an end, then its entropy increases, its structures and functions are violated, and the body disintegrates and dies.

It is important to note that all the processes of transformation and utilization of energy are strictly ordered and organized by structural elements of the phenotype—enzymes, organelles, and membranes. Their appearance and functioning is precisely controlled by the genome. So once again the genome is the central figure of global movement, which directs energy processes for its own satisfaction.

## 17.4  Thermodynamic Mechanisms

The life of individuals, their metabolism, and the maintenance of order are based on anabolic anti-entropic processes for overcoming chaos that arise spontaneously in biological systems. Any work, and in particular within biological systems, is connected with the transformation of energy. The study of energy and the mechanisms underlying its transformations in various systems constitutes the science of thermodynamics.

Living systems are nonequilibrium systems with a molecular structural basis and unlimited heterogeneity. Nonequilibrium systems are ones that can change their specific parameters either independently or under the impact of some forces. If the parameters of a nonequilibrium system do not change with time, this state of the system is said to be stationary. In other words, living bodies are nonequilibrium and stationary at the same time, being able to work towards the maintenance of their dynamic stability at any moment. In various parts of the system, the meaning of the parameters may differ significantly, e.g., different concentrations of molecules in a cell. In these systems, gradients of certain parameters are maintained and various processes take place, such as diffusion, osmosis, biochemical reactions, etc., which are characteristic of living bodies. Therefore, a mandatory precondition for the ability of any thermodynamic system to do work is the existence of some differences between different points of the system. Such a dynamic condition is maintained by means of the flows of matter and energy that pass through it. Systems that exchange matter and energy with the environment are called open systems. Thus, living bodies are open and nonequilibrium, but stable systems.

One significant characteristic of systems is their internal energy, which is the sum total of kinetic and potential energies contained in its elements. Internal energy is a function of the condition of a system and has a particular value for a given state. Living systems have very high internal energy, since they are highly organized structures. There are two main forms of energy transmission: heat (Q)— energy transmission according to temperature gradient in the form of matter with disordered motion, and work (A)—energy transmission in the form of ordered motion, related to the motion of objects.

There are two main laws of thermodynamics:

1. The first law of thermodynamics or the law of conservation and transformation of energy says that energy can neither be created nor destroyed. It can only change forms. An example of this is photosynthesis, wherein photon energy is

transformed into the energy of chemical bonds in organic matter. The first law of thermodynamics is a general law of energy conservation. However, it does not define the ability of any particular energy to transform and does not indicate its directionality.

2. The second law of thermodynamics, or the law of entropy, says that the only processes that can develop independently are ones in which the system goes from fewer possible to more possible states. According to this, spontaneous processes (without energy costs) move toward increasing chaos, i.e., the entropy of isolated systems increases gradually and irreversibly. This happens because, in the processes of movement and work, when energy is transformed from one form to another, the amount of free energy of a system decreases, since some of it is disseminated. The reverse processes are impossible without an additional supply of energy. In other words, a certain amount of energy should be used in order to go from chaos to order.

The laws of thermodynamics are applicable to living bodies as well, since their existence is based on the thermal motion of molecules and performance of work. Many processes, during which energy goes from one form to another, take place in living bodies all the time. For example, the chemical energy released by hydrolysis of ATP is transformed into kinetic energy of molecular migration through a membrane, although some energy is dissipated as heat. During oxidation of glucoses, only 55 % of the energy obtained is stored, while the rest goes into uncontrollable thermal motion. Chaotic thermal motion of cytoplasmic molecules provides energy for the processes of diffusion, osmosis, and interactions between molecules. Enzymes are also unique protein molecules that exploit the chaotic motions of surrounding molecules for their work, as well as the energy of thermal oscillations and fluctuations within their own molecule. These are micromechanical constructions, whose parts move in relation to each other under the impact of thermal fluctuations. The motions of parts of enzyme molecules are selective, so for example, specific segments of the active center ensure highly precise work in processing substrates at the expense of internal heat. In this way, some of the disseminated energy can be turned into work under suitable conditions.

However, the uncontrollable thermal motions of molecules initiate processes that increase the entropy. Many molecules have very high speeds, and possess high enough kinetic energies to inflict uncontrollable interactions and destruction of order in biological structures. Elements of highly organized biosystems are also destroyed under the impact of radicals, radiation, etc. However, living organisms can remain intact for rather long periods. In other words, it is almost as though they are able to disobey the second law of thermodynamics for a certain period of time. This happens because, in spite of their autonomy, living bodies are not isolated, i.e., they constantly exchange energy and matter with the external environment. Consumed and then transformed matter and energy constantly serve to counteract fluctuations and disturbances in biological systems.

A living body can be imagined as a conflicting biochemical system which can only function in the temperature range 0–40 °C (though some thermophilic bacteria are known to function even above 100 °C), and whose work consists of the constant removal of structural defects that appear incessantly at these temperatures as a result of thermal motions of molecules. In other words, living processes are possible only within a narrow range of rather low temperatures, but these temperatures are nevertheless rather high for living matter, tending to produce dissipation processes. The high speeds of chaotic motion of high energy molecules cause unavoidable thermal destruction of thermolabile structures in living bodies. Indeed, macromolecules and their complexes can be damaged, along with structural proteins, enzymes, nucleic acids, ribosomes, membranes, and many others. The result is the destruction of order in the cellular system, i.e., an increase in its entropy. Thus, living bodies are dissipating; they are in a state of constant destruction. Only the presence of ongoing anabolic restoration processes (for example, synthesis of proteins, DNA repair) can decrease the level of entropy and therefore ensure the relatively stable existence of biosystems. This internal conflict or duality of existence of living systems (isochronous destruction-restoration) is one of the main qualities that distinguishes living bodies from non-living ones. Non-biological systems constantly increase their entropy, whereas biosystems always try to maintain order.

In addition, in contrast to other systems, biological systems are not only relatively stable nonequilibrium formations, but they also have the property of being able to complicate their structural and functional characteristics by exploiting information during development. In other words, they have a tendency slow down the emergence of entropy by consuming energy and exploiting information.

Any artificially created mechanical system does not have to work and will not lose its structure. In contrast, a living body must function all the time—it is an active and constantly running system. If, for some reason, an organism stops fulfilling its main functions at the vital temperature, it will lose its structure irreversibly and die. All the processes of metabolism in a living system are catalytic, the catalysts being represented by special proteins called enzymes. It is only through the catalytic nature of internal processes that a controlled transformation of chemical energy from food products into the necessary work can take place at a high enough rate at the relatively low temperatures of existence of biological systems.

Biological dissipative systems maintain their orderliness only by virtue of anabolic processes that take place all the time to ensure construction or restoration of highly organized structure, which is continually subject to thermodynamic destruction. However, if biological objects are cooled down or frozen, the processes of thermal destruction and metabolic processes slow down simultaneously. That is why, during a prolonged period, no destruction of the organism will take place. In other words, anabiosis occurs. Such a preserved system may revive when thawed, unless its structure has been significantly damaged. Thus, ancient microorganisms revive in laboratories after many thousands of years when they are retrieved from permafrost, where they have been hiding from the laws of thermodynamics.

The organization and order of living bodies is maintained at the expense of mechanisms coupling energy flows in cells, where energy produced in the

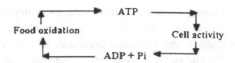

**Fig. 17.3** The ATP-ADP cycle is the main mechanism for conjugation of incoming energy and cell activity. This 'simple' mechanism, realized by special enzymes, is typical to virtually all living bodies (the only known exceptions among cell forms that do not produce their own ATP are rickettsia and clamydia). It is a biochemical mechanism conjugating endothermic and exothermic reactions, i.e., disintegration with creation

processes of destruction is used to sustain restoration reactions. Life is maintained by many anabolic chemical reactions which cannot develop independently outside the biological system, since it would be thermodynamically impossible, e.g., synthesis of proteins, nucleic acids, ATP, etc. However, they are carried out in cells on a permanent basis and at high rates. This happens because they couple with exothermic processes that can develop with energy release, e.g., ATP hydrolysis. The resulting energy is used by the cell to carry out various tasks. As one of the most energy consuming molecules, ATP plays an important role in the processes of coupled reactions (Fig. 17.3). During ATP hydrolysis, a significant amount of energy is released for use in anabolic reactions. Reciprocal ATP-ADP transformations are used by an organism as a key mechanism for coupling alternate thermodynamic flows of matter and energy. It is also important to emphasize that the coupling of energetically advantageous and disadvantageous reactions is carried out only with the participation of coupling enzymes.

It is very important that living bodies should be able to use spontaneous internal processes seeking to increase entropy in order to perform useful work, such as diffusion, osmosis, charged particle transport, reduction–oxidation, and so on. In other words, the structural order (negentropy) and internal energy of molecules (enthalpy) of biological systems possess some potential to perform work. Consequently, spontaneous thermal destruction is not only a force of dissipation, but also a *force* provoking restoration and maintenance of standards of organization in living bodies (along with use of energy and matter from the environment). Processes realizing such potential in biosystems bring about enzyme controlled biocreation and restoration of the system.

It may be some thermodynamic mechanism of life, an uncontrollable spontaneous flow of destruction, that sets in motion a controlled flow of system restoration. The mechanism described here is very significant for life and development, because it becomes clear that neither expenditure of matter and energy, nor any information are required to remove the old and unnecessary—all organized systems can be destroyed spontaneously. Energy, matter, and information are only used at the creation stage.

The same thermodynamic principle is also used by Nature for the global process of evolution. Billions of individuals of millions of species are destroyed spontaneously all the time, in compliance with the law of entropy. However, the mechanism of self-reproduction prevents these species from disappearing from the Earth. In the process of reproduction, genomes of living bodies undergo mutations and recombination, whence some modified creatures are restored with evolutionarily favorable traits. The fate of these systems is also spontaneous destruction, but with controlled restoration of new systems based on the slightly modified programs. And so it goes on for billions of years and generations.

In short, living organisms are a type of open nonequilibrium dissipative system, working incessantly against their own destruction, following the genetic information they enclose. Under the impact of thermal motions of molecules, the structure of rather thermolabile elements of biosystems is constantly and uncontrollably destroyed. At the same time, controllable anabolic processes also continually restore the highly informative structure of living systems. In other words, life and metabolism are underpinned by processes able to overcome, in a controllable way, the thermal chaos that spontaneously appears in biological systems at the vital temperature. It is the very vector of spontaneous thermal destruction that constantly creates multiple changes of state in the millions of micro- and macro-systems making up a cell. This maintains its nonequilibrium state and sets up conditions for the constant controlled flow of matter and energy that can create and maintain order.

We live in a decaying world. All things said, it is the genome that fights against the second law of thermodynamics in Nature. It does this by organizing structures and processes that undo the consequences of thermal destruction. This great supervisor cannot prevent the unavoidable, but it can successfully restore itself and its phenotypical surroundings using the very same thermal potential. Controlled and spontaneous destruction is the basis of biocreation!

## 17.5 Electrostatic and Electrodynamic Mechanisms

The content of a cell is a colloid. It comprises approximately 70 % of water, in which various, mostly organic molecules are dissolved. Life is maintained by means of interactions between these molecules. Such interactions are possible only in the liquid environment, where the high speed of their movement can be maintained. Both organic and non-organic molecules are in a dissociated condition, i.e., they carry a system of electric charges on their surfaces. The water molecule is also a dipole that carries both positive and negative charges. It is thus clear that practically all the content of cells is represented by charged molecules. Therefore, cells and multicellular organisms incorporate many processes involving electric and electromagnetic interactions.

Living bodies, cells, and the intercellular contents can conduct electricity. Electrical conductivity is the ability of bodies or substances to conduct electric current. In particular, electrical conductivity plays a significant role in the functions of electroexcitable tissues: nerves and muscle. Electrical properties of living bodies are significantly more complex than those of non-living objects, since an organism is a heterogeneous set of ions with variable concentration in space and time. The other peculiarity of electrical conductivity in living systems is the transfer of substances along with electric charges. This property assists in various directed mass transfer events in cells. For example, all polar and ionized substances are transferred from one compartment to another through a branched system of biomembranes by means of electrical interactions.

The process of dissolving molecules in the cytoplasm and then maintaining that dissolved state is governed by the electrical interactions of those molecules with aqueous dipoles. In other words, the state of the internal contents peculiar to cells is an ionized state of electrodynamic nature. This is a heterogeneous system of moving, polarized, interacting molecules with positive or negative charges. Many biochemical reactions occur through electrical interactions. For example, catalysis is the main mechanism of life and it is based on the electrical interactions in the active centers of enzymes. Acid–base and covalent catalyses are distinguished. In the first case, enzyme activity is conditioned by the participation of amino acid residues of an active center in chemical reactions that have functional groups of donors and/or acceptors of protons. Their electrostatic interactions with the substrate condition the redistribution of charges, weakening of chemical bonds, and removal and transfer of specific charged functional groups or parts of molecules. Covalent catalysis is also connected with electrostatic interactions between the substrate and charged groups of the active center, during which temporary covalent connections are formed.

All cells are enclosed in membranes consisting of phospholipid and protein molecules. Both phospholipids and proteins possess amphipolar properties. That is, hydrophobic parts of molecules are located inside the membrane (without a charge), while hydrophilic parts are situated on the surface. This creates a specific electric field around cells. In fact, it is an electromagnetic field, since the charges on the surface of cells and the anions and cations connected with them can move around. Cell surfaces are normally negatively charged, while the cytoplasm is positively charged.

Since biomembranes possess selective permeability, various concentrations of ions such as $K^+$, $Na^+$, $Ca^{++}$, $Cl^-$, $H^+$, $OH^-$, etc., on either side of a membrane create the membrane electric potential. This rest potential of cellular membranes may reach up to 50–100 mV. The main contribution to the value of the rest potential comes from sodium, potassium, and chlorine ions. This diffusive potential is the driving force for many vital processes. For example, many molecules, such as amino acids, glucose, urine, water, $K^+$, $Na^+$, etc., are introduced or taken out of the cells by means of special carriers exploiting the electrochemical gradient.

Some tissues are said to be electroexcitable. These are tissues in which the cell membranes have high electrical rest potential and possess the ability to reversibly depolarize. In such electroexcitable cells of nerves, muscle systems, and the heart, an action potential (110–120 mV) can arise which is conditioned by directed and controllable flows of ions through a membrane under the influence of electrical or chemical signals. The distribution of this potential, e.g., along the axon, causes the conduction of an electrical impulse. Cells of the myocardium have gap junctions between them, through which excitability rapidly spreads throughout the whole organ in the form of an electric current. This causes practically simultaneous contraction of the whole myocardium.

Electrical phenomena also underlie perception and dissemination of information in living bodies. In particular, all signals from the external and internal environment are transformed by receptors into electric current, which is distributed in a targeted way through special cellular structures at large distances, where it has its effects.

It is also known that the work of the brain is based on the electrical activity of billions of interconnected neurons. Electric currents circulate between them through billions of branches, exploiting still poorly understood phenomena of perception, recording, analysis, and use of information received through thousands of channels from hundreds of different receptors. It is interesting that irritants of any kind are transformed by the receptors into specific electric signals, which are transferred to the brain (central computer) via electrical networks, where they are processed. The external manifestation of electrical activity in the heart or brain is electromagnetic oscillations, which can be registered with an electrocardiograph or encephalograph, respectively.

The key mechanism of life is a set of energy transformation processes in cells. It is interesting that electrodynamic mechanisms are more effective than chemical ones. For example, absorption of electromagnetic photon energy by chlorophyll underlies photosynthesis. Then, as a result of certain physicochemical processes, the concentration of protons ($H^+$) and electrons ($e^-$) increases in the thylakoid matrix, and this creates an electrochemical potential on the membrane (Fig. 17.1). High-energy electrons are transported by special proteins through the transport chain located inside the thylakoid membrane (constituting an electric current). The electrons subsequently release their energy, which is used for transfer of protons through a membrane into the thylakoid matrix. This raises the concentration of protons on one side of the membrane and increases the electrochemical membrane potential. The energy of this high proton gradient is then used for ATP synthesis. Thylakoid membranes contain special molecular complexes called $H^+$-ATP synthases. The energy of a proton moving through them is transformed into the energy of chemical bonds in ATP synthesis. In other words, through a number of electrodynamic and electrostatic processes, chloroplasts transform the energy of electromagnetic radiation into proton potential energy on biomembranes, and this is used to transform and store energy in the chemical bonds of ATP.

Other energy converters are mitochondria. These oxidize various organic molecules, resulting in the creation of a significant amount of protons and electrons. Protein-lipid complexes in the respiratory chain of internal membranes of mitochondria contain components that have different redox potentials. This assists the spontaneous movement of electrons along the components of the respiratory chain from the high electrochemical potential to the low one. The energy of this electric current is used to transport protons through a membrane and create a powerful proton gradient (Fig. 17.2). The energy of this potential, as on the thylakoid membranes, is then used to transfer protons through ATP-synthases and hence transform electrokinetic energy into the energy of chemical ATP bonds.

All biochemical processes develop in a liquid environment. The acidity (pH) of the internal environment of cells, the intracellular environment, and the fluids of multicellular organisms is close to neutral, though it has some peculiarities in various bio-objects. Enzymatic processes are only possible at optimal pH values. It is known that the acidity is conditioned by the qualitative content of hydrogen ions ($H^+$). Even slight changes in pH (alkalosis and acidosis) can perturb biochemical processes, metabolism, and functions. To maintain the stability of the acidity of the internal environment, biosystems have buffer systems (bipolar charged systems) that maintain the required proton concentration.

Thus, cells are complex microheterogeneous systems of charged elements. Practically all the processes connected with the existence of living bodies are conditioned and accompanied by electrochemical phenomena. All the particles of a cell interact electrically with one another. One can assume the presence of a common electromagnetic cellular field that connects all electrical units into an integrated system. It is clear that any perturbation of the electrical equilibrium, e.g., breaks in chemical bonds, should cause fluctuations in the surrounding molecules. This will in turn lead to a change in the general condition of the components of the system and the system as whole. Thus, electrical interactions can constitute a finely tuned mechanism for integrating and regulating cells and cell systems. If we add to this the hypothetical electromagnetic field of a genome, this can also provide a mechanism for its global impact on any part of the cell, any molecules, and any structure. The reaction of elements of the system and their reverse connection can be just as quick.

Thus, in cells, there are global electrical processes and interactions, and electric currents flowing in thousands of ways and in all directions. Millions of chemical reactions, which we judge by external manifestations and the formation of various kinds of matter, are actually of an electrical nature. All chemical transformations are connected with electron and proton interactions, shifts, and movements of electric charges. This is not a chaotic movement, but controlled and coordinated transfer of electric charges from one molecule to another, from one organelle to another, and from one part of a cell to another. Through this connection, the coordination between different parts of a cell takes place at colossal speeds, comparable with the speed of an electric current. In an instant of time, even the most remote and minimal portions of a cell can be involved and react. In this way, a cell or a genome can control the presence of specific currents and their directions. Electrical processes integrate all the multiheterogeneous components of a cell into a single system. Thus, cells can be imagined as microelectronic physicochemical machines based on electronic connections, retaining their integrity by means of electromagnetic interactions.

## 17.6  Quantum Mechanisms

The majority of biological processes and mechanisms are conditioned by quantum mechanical interactions of molecules and elementary particles. Quantum mechanics is the science that studies laws of movement and interaction of microparticles (electrons, photons, protons, neutrons, etc.), as well as their systems. A quantum is an amount of particles, namely the minimum amount needed to possess a specific property of matter. A particle of an electromagnetic field is a photon; a particle of a gravitational field is a graviton, etc. A quantum is a minimal amount, always discrete, by which a physical quantity can change (mass, energy, action, momentum, etc.). Quantization is effectively discretization, or division of some physical quantity into discrete portions.

All material bodies, either living or non-living, consist of atoms and molecules. An atom is a quantum–mechanical particle that consists of a positively charged nucleus and negatively charged electrons distributed along various orbits of rotation. A chemical element is a specific type of atom with the given electrical charge on its nucleus. Atoms of the majority of elements possess the ability to give away or bond with electrons. If an atom has empty electronic orbits, it is unstable. It can easily participate in chemical reactions by giving away or gaining electrons. In other words, the ability of elements to react is determined by peculiarities in the structure of the external electronic shells of atoms.

Atoms form molecules of various kinds by binding together with chemical bonds. A chemical bond is a stable interaction of atoms through their electrons, leading to the formation of multi-atomic chemical compounds or molecules. Combinations of dozens of different atoms can form a multitude of molecules with specific molecular masses and various configurations. A molecule is the smallest particle of matter that determines its physical and chemical properties. After the formation of a chemical bond, atoms lose their individuality, and the properties of molecules differ from those of the elements that compose it. The structure and properties of molecules are determined by the spatial and energetic order of the quantum–mechanical system formed by atoms and electrons. Chemical reactions, including biochemical reactions, consist in a transformation of one or several molecules into others that differ by their composition, structure, or properties. During a reaction process, the total number of atoms and elementary particles does not change. In other words, chemical reactions are quantum processes involving only the redistribution of electrons and regrouping of nuclei, whereas the nuclei of the atoms remain unchanged.

Atoms and molecules of cells are in constant thermal motion and can collide with each other many times. When molecules collide, sufficient energy may be released to change chemical bonds, causing disruption, transformations, or the formation of new ones. Thus, new bonds of atoms and new molecules are formed on the basis of molecular and quantum processes.

Chemical reactions can be reversible or irreversible. Spontaneously, reactions move towards decreased energy and increased entropy in a system. The possibility of the development of reactions and their rates depend on a number of conditions. The process is affected by the temperature, pressure, mechanical impacts, electric currents, catalysts, etc. Reactions can be managed by changing these conditions. Complex compounds of changeable composition (in particular, many organic molecules) in which the bonds between groups of atoms can be weakened are the ones that depend most on this kind of factor.

A group of reacting molecules composes a chemical system, which can be balanced or nonequilibrium. In balanced systems, reversible reactions take place, and in nonequilibrium systems, irreversible chain or branching reactions occur. Cells are nonequilibrium biochemical systems. These are systems in which fluctuations, instability, and lack of specificity in the development of processes can arise. However, the directionality and intensiveness of biochemical processes in cells is strictly regulated by selective catalysis. A chemical process is a sequence of reactions, leading through a number of intermediary stages to the required form of matter as a result of changes in the chemical conditions of the system. The cascade of chemical transformations consists of sequential processes of redistribution of electrons and atoms. Thus, it is obvious that the bases for chemical and therefore biochemical transformations are quantum modifications.

Practically all metabolic and physiological processes are connected with movements of elementary particles. For example, these include oxidizing and reducing reactions, photosynthesis, oxidative phosphorylation, polarization and depolarization of membranes, the phenomenon of sight, conductivity of a nerve impulse, electrical activity in the brain, etc.

Now, let us consider some examples of specific quantum-biological processes. The main source of energy of vital processes on Earth is the Sun. Photosynthesis serves as the main energy transformation process from the flow of photons into the energy of chemical bonds in ATP, and then chemical bonds in organic molecules. This is a characteristic example of quantum processes that form the basis of life.

The essence of the quantum-biological transformations in the first phase of photosynthesis is the absorption of quanta of light energy and their transformation into the energy of chemical bonds in ATP in a number of stages. Absorption of light is achieved by chlorophylls. These complex organic molecules are contained in special photosynthesizing structures or organelles of plant cells called chloroplasts. They have special complex membrane structures—thylakoids. Chlorophyll molecules are built into the thylakoid membrane itself, where they capture photons of light.

At least 5 forms of chlorophyll are known (**a, b, c1, c2,** and **d**). However, in a particular organism, chlorophyll usually exists in just one (rarely) or two forms. The first form, which is inherent to all plants and cyanobacteria, is type **a**. It directly absorbs the photon of light with the wavelength of 700 nm and easily changes to the active state by losing the electron which has absorbed the energy quantum. Chlorophyll of the second type (**b** for higher plants, **c1** and **c2** for the majority of algae, and **d** for red algae) absorbs photons at 680 nm and uses its energy for photooxidation of water by the protein Fe–S.

Photolysis of water produces free protons, electrons, and oxygen. The concentration of protons and electrons in the thylakoid matrix increases, and an electrochemical potential is created on the membrane. Highly energetic electrons of both types of chlorophylls are transported along the thylakoid membrane by special proteins. The electrons thereby lose energy, which is used by other proteins to transfer protons through a membrane into the thylakoid matrix. This causes an even higher concentration of protons and increases the electrochemical gradient through the membrane. The energy stored in this high proton gradient is used for ATP synthesis. Thylakoid membranes contain special $H^+$-ATP synthases. The energy of the concentrated protons that move across this gradient is transformed in the active center of these enzymes into chemical bond energy by synthesizing ATP molecules. Thus, the essence of the light phase of photosynthesis is dynamics and sequential transformations of elementary particles—photons, protons, and electrons, i.e., quantum–mechanical processes that eventually lead to the formation of macro-energetic ATP bonds.

In the same manner, the processes of oxidative phosphorylation that develop in the mitochondria of all organisms have a quantum basis. In the matrix of mitochondria, enzymatic oxidation of organic matter takes place by removal of hydrogen ions (protons) and electrons. Electrons are carried by enzymes to the respiratory chain where, moving from one element of the chain to another, they gradually release the energy. The latter is immediately used to transport protons from a matrix to an intermembrane space. In this way, a membrane electrochemical gradient of protons with high potential energy can be created. An internal membrane of mitochondria contains integral molecular complexes of $H^+$-ATP synthase. Passing through such a membrane, protons can gradually give away their energy, and this is also transformed into the energy of chemical bonds of ATP in a quantum way.

The phenomenon of selective enzyme catalysis is the main mechanism for carrying out all biochemical reactions and processes, functions, and structure formations, i.e., it lies at the very foundation of life. The work of the tens of thousands of different enzymes is based on quantum–mechanical processes. In the active center of enzymes, there are functional atoms or their groups that can connect and orient the substrates of reactions. This happens through the formation of various chemical bonds. Any chemical bond consists of interacting electrons, or electrons and protons. Other functional groups of the active center affect a strictly defined chemical bond through their electrons and protons. Such an impact causes displacements of electrons and protons, changes in the conformation of substrates, and the weakening of bonds to the point where they are broken and electron orbitals gain other orientations.

However, not much is yet known about the abilities of remote quantum interactions of molecules and cells through various fields and radiation ("wireless" transfer of energy and information). This may be a promising direction for further research.

Thus, non-specific quantum mechanisms of interaction that are characteristic of matter are the basis for the formation of all biological structures and the development of vital processes. The peculiarity of these processes in living organisms is connected with the creation and maintenance of specific conditions (based on genetic information) for the development of quantum processes only in specific places at specific times, with specific molecules, and in a specific direction. In other words, quantum mechanisms are the main mechanisms used by living systems, but only with a view to creating order in cell structures and processes, as programmed by a genome.

# Chapter 18
# Cell Mechanisms

## 18.1 Cytological and Cytogenetic Processes and Mechanisms

Not only do molecular microprocesses go on in cells, but so also do many kinds of purposeful cytological and cytogenetic macroprocesses, which are related to interaction, transformation, and controlled movement of colossal organized cellular masses. Cells consist of many complex interrelated and interacting parts. Discreteness provides inner movement and interaction of cell components, and therefore manifestation of their various properties and functions. Naturally determined location and controlled interaction of elements of the system conditions the emergence of qualitatively new properties and characteristics of cells.

Such huge sophisticated complexes of interacting organelles and cell parts (e.g., the vesicular transport system) exist and act as an integrated device, indeed as integrated cytosystems. Moreover, they happen to be interacting and interrelating with all the other cellular macrostructures. It is their well-coordinated interaction that provides the life and functioning of cells as integrated autonomous bodies. It is important that all organelles are 'immersed' in and interact with the highly organized and complex cytosol. Mitochondria, lysosomes, membranes and vesicles of the endoplasmic reticulum and Golgi apparatus, ribosomes, chromosomes, and others constantly move, change shape, size, and structural and functional state. All of this occurs simultaneously and interconnectedly, in a single medium and a single operative space.

One of the most expressive and complicated examples of macroprocesses is the phenomenon of cell division, which is connected with the ordered rearrangement of huge cellular masses, structures, and organelles. Let us take a look at the processes connected with cell division from the point of view of the mechanisms underlying the transformation and relocation of intracellular masses (Fig. 18.1).

The cell cycle is characterized by numerous processes that occur in the cell: growth, differentiation, functioning, etc. It consists of a prolonged period of interphase and short periods of mitosis and cytokinesis.

G. Zhegunov, *The Dual Nature of Life*, The Frontiers Collection,
DOI: 10.1007/978-3-642-30394-4_18, © Springer-Verlag Berlin Heidelberg 2012

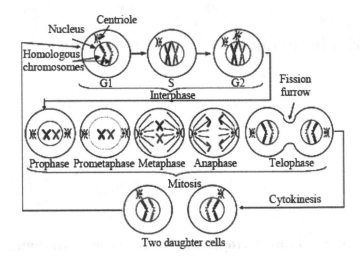

**Fig. 18.1** The main stages in the transformation of genetic material during the cell cycle as an example of a macroprocess. See text for explanation. From the genomic perspective, cell division is just a mechanism for copying and replication. The diagram shows how the genome uses the phenotypic framework for its selfish ends

The first period of interphase is presynthetic ($G_1$). During this period, the cell and its genome function actively, with DNA managing RNA and protein synthesis. The nuclei of such cells contain the diploid number of chromosomes, each of which is represented by one DNA molecule.

During the next synthetic (S) period, DNA is synthesized and duplicated. As a result, every chromosome consists of two daughter molecules of DNA, which are connected together at the centromere. The number of genes thus doubles. The amount of chromatin proteins also doubles. DNA replication is a very important moment in the preparation for cell division. Replication is the basis for both nonsexual and sexual reproduction, thereby ensuring the continuity of life.

During the postsynthetic period ($G_2$), cells prepare for the division of genetic material—mitosis. Spiralization and condensation of chromatin begin, along with the gradual destruction of the cytoskeleton. The synthesis of ATP, proteins, RNA, lipids, and carbohydrates increases. New elements of the cell are formed.

The described order of events in the interphase gives evidence of the many complicated structure-functional interactions of molecules and parts of cells, and the significant rearrangements and relocations occurring in the preparation for division.

*Mitosis* is a complicated mechanism that ensures the division of genetic material into portions after replication in such a way that each of two daughter nuclei receives a complete copy of the genetic information. As a result, the nuclei of all cells in the organism contain a qualitatively and quantitatively equal set of chromosomes. Mitosis is supported not only by the action of intranuclear forces, but also by the work of all intracellular structures. Mitosis is subdivided into five consecutive phases: prophase, prometaphase, metaphase, anaphase, and telophase.

*Prophase.* When the prophase starts, the nuclear material undergoes significant modifications. Long fibers of chromatin become shorter, concentrate, and form loops and spirals. They can be clearly seen with a light microscope as separate chromosome bodies. At this stage each chromosome consists of two chromatids, which are located close to each other throughout almost their whole length. Each chromatid consists of one molecule of DNA wrapped around histone proteins. At the area of tight contact between the chromatids, there is a special shared DNA DNA region called the centromere, which will later become connected with the spindle. Pairs of centrioles move away from each other towards the opposite ends of a cell, forming two division poles. These macrostructures are involved in organizing microtubules of spindle formed from proteins of the ruined cytoskeleton.

*Prometaphase.* The beginning of the prometaphase is characterized by fast disintegration of the nuclear membrane into small vesicles. Caryoplasm is mixed with cytoplasm, forming mixoplasm, and microtubules of spindle can now reach the chromosomes. The latter have become even more dense and special protein elements called kinetochores (from which microtubules begin) are formed on the centromeres. Groups of microtubules of the maturation spindle interact with microtubules of the kinetochores causing the chromosomes to move.

*Metaphase.* Concentration of chromosomes in the equatorial area indicates that the cell has reached the metaphase. Grouped in this way, chromosomes are referred to as the metaphase plate, to which they are fixed by microtubular tension. The microtubules attached to kinetochores then start to pull chromosomes in opposite directions in such a way that opposite chromatids move away from each other. During the metaphase, the chromosome structure is clear and can be viewed under the microscope. At this stage, each chromosome consists of two chromatids which have already diverged at the ends, whence the chromosomes have an X-shape. At the end of this phase, replication of the centromere segment of DNA is finished and the chromatids are completely disconnected.

*Anaphase.* During the anaphase, chromatids of each chromosome migrate towards opposite ends of the cell. Their separation occurs simultaneously and rapidly. All chromatids move with the same approximate speed of around 1 μm/min. Each anaphase chromosome contains one molecule of DNA. They have a rod-like shape, bent at the centromere. At the end of the anaphase, two completely equivalent sets of chromosomes are gathered in different parts of the dividing cell.

*Telophase.* Two identical numbers of chromosomes are located at the opposite poles of a cell. The spindle has disintegrated. Around each group of chromosomes new nuclear membranes are formed by fusion of vesicles. As a result, typical nuclei are formed. The hereditary material of the chromosomes begins to unfold to reach the typical chromatin condition for the interphase. Nucleoli reappear. When these modifications are completed, mitosis comes to an end and each daughter nucleus enters the next cycle. During the first phases of mitosis, large membrane organelles like the Golgi apparatus, endoplasmic reticulum, and nuclear membrane are first separated into smaller fragments and vesicles. This ensures their equal distribution between the daughter cells during the telophase.

The result of mitosis is the formation of two nuclei, which are genetically identical to each other and have a complete number of DNA molecules, as required to relay all genetic information. In this way, the intracellular migration of matter and distribution of genetic material are achieved for millions of years, with complete preservation of the structure and individuality of every chromosome in a complicated cell cycle and in all ensuing processes.

*Cytokinesis* is the process of mechanical division of a mother cell into two daughter cells. This process occurs differently for various groups of organisms: either from the 'inside' by formation of the phragmoplast (a vacuole platelet, which splits the cells), or from the 'outside' by means of constriction. The latter is typical for animals and involves the following stages: (1) a cleavage furrow is formed at the cell equator from microfilaments of cytoskeleton that make up a contractile ring; (2) the ring progressively contracts and the furrow deepens more and more along the perimeter; (3) after a certain time, the maternal cell divides into two daughter cells.

In some cases a programmed unequal distribution of cytoplasm components may occur during cytokinesis. In particular, a cell may divide to form daughter cells that are uneven in size. For example oocytes and polar bodies are formed in this way in the process of oogenesis. Alternatively, before cytokinesis, some components of the cytoplasm may concentrate in a certain part of the cell and be passed into only one of the two daughter cells.

It is thus clear that mitosis and cytokinesis are very complex stages in the life of a cell, consisting of many cytological and cytogenetic processes. These processes are connected, not only with molecular mechanisms and interactions, but also with many processes of nonrandom transformation, reorganization, and migration of significant ordered amounts of material inside the cell. The main intracellular macroprocesses during mitosis are:

1  Formation of complicated macroscopic complexes—chromosomes, from fibers of chromatin.
2  Formation of two dividing poles and formation of microtubule spindle system—generation of absolutely new global macrostructure.
3  Disintegration of nucleus and its membrane. Disassembly into particles and formation of special membrane vesicles. A unique method of conservation and storage of parts and the content of the nucleus. A huge, highly organized macrostructure that occupied up to 50 % of the cell volume ceases to exist.
4  Formation of complex contents of a cell—mixoplasm, no less ordered and organized structure than cytosol or karyoplasm.
5  Controlled and purposeful growth of microtubules of spindle—massive synthesis of proteins and controlled assembly of mitotic apparatus.
6  Precise migration and interaction of dozens of chromosomes and hundreds of microtubules.
7  Organized and purposeful movement of chromosome bodies through viscous matrix of cytosol and accurate orientation of them in the central part of a cell.
8  Simultaneous division of chromosome complex into chromatids.

9 Controlled movement of chromatid bodies through dense matrix towards opposite poles of the cell.

10 Organized disintegration and utilization of large amount of spindle microtubules.

11 Fusion of nuclear membrane vesicles and restoration of membranes and contents of nucleus around daughter sets of chromosomes.

12 Unfolding of chromosomes and formation of a network of functioning chromatin.

13 Formation and deepening of the division furrow of the maternal cell. Organized and purposeful redistribution of complexes and organelles. Division of maternal cell body into 2 daughter cells.

So we see that, not only do molecular metabolic processes (not mentioned here) occur in cells, but so also do all kinds of possible macroprocesses, connected with interaction, reorganization, and migration of colossal organized cell masses, large parts of a cell, its organelles, and large molecular complexes. Moreover, all these sophisticated processes are implemented very quickly and precisely by the cell with a high level of organization and in strict order. Such processes are repeated infinitely often with almost no errors or failures.

Transformations of cumbersome genetic material and the accompanying cytological mechanisms excite admiration by their organization, accuracy, reasonableness, economy, and expediency. Interactions and transformations of other intracellular macrostructures are refined and accurate as well (e.g., assembly and disassembly of cytoskeleton elements, formation and migration of membrane vesicles and organelles, division and migration of mitochondria, etc.). Expedient interactions of various parts of a cell and precise and purposeful transformations of large molecular structures and complexes are implemented on the basis of complicated cytological macromechanisms and processes, which maintain its macrostructural and functional order.

Thus, along with numerous molecular microprocesses (e.g., enzymatic catalysis, molecular synthesis, biological oxidation, etc.), there are many macroprocesses and mechanisms connected with interaction, metamorphosis, and migration of organized cellular macrostructures.

However the nature of causality, congruence, and precision of migration, recognition, and interaction of complicated molecular processes as a comprehensive whole is not completely understood yet. Particular biochemical reactions, such as transformations of certain molecules in active centers of enzymes, are now quite well understood, as are the mechanisms and forces of metabolic processes, which are quite well explained using the laws of physics and chemistry. On the other hand, it remains almost impossible to understand the complexity of patterns and the repeated accuracy of transformations of cellular macrostructures. We still do not understand what causes the organization of their migrations and interactions, expediency, interrelations, and controllability.

## 18.2  Functional Systems of Cells

Various physiological systems of multicellular organisms are known. They are complexes of organs and tissues that fulfill certain functions for the purposes of the body as a whole. In particular, systems like the neural, endocrine, cardiovascular, excretory, respiratory, reproductive, immune, and others have been well studied.

On the basis of components of physiological systems, and as the need arises, multicellular organisms form even more complicated dynamic associations, or functional systems. These systems are self-organizing and self-regulating dynamic formations with components whose activity conditions the achievement of adaptive results for the whole organism. They perform a complex of functions aimed at ensuring global functions, such as survival, reproduction, and dispersion (e.g., the integration system of the organism, the sensor system, the system for maintenance of homeostasis, the energy transformation system, the system for protection from environmental factors, the reproductive system, the system for maintenance of integrity, the thermoregulation system, etc.). The segregation of definite functional systems is quite relative, since the composition of elements of functional systems may vary significantly depending on the tasks of an organism. For example, the global task of integration of the organism is achieved in the first place by interrelated operation of the neural, endocrine, and immune systems. Together they form a qualitatively new functional system that unites absolutely all cells and intercellular substances into an integrated organism. The functional system of thermoregulation of a body includes integument tissues, the cardio-vascular system, certain endocrine organs, certain properties of skeletal muscles, certain processes of cellular metabolism, etc.

In a similar way, cells are also able to form certain groups of interrelated elements, organelles, and compartments, depending on their needs. This is achieved by the block mode of organization and functioning of cells.

Major functional systems of cells may be defined on the basis of the performed functions. They are formed by directed interactions between organelles, cellular parts, and molecular complexes at the moment of carrying out a certain task, and act for the purposes of the cellular functions as a whole. The goals of these systems may match with those of corresponding systems of multicellular organisms, or may be typical only for the cell itself. They may express peculiarities of separate cells or of cells in a system of tissues and organs.

Listed here are examples of a series of functional systems, typical for most cells: (1) System of nourishment. (2) System of substance transformation. (3) Energy transformation system. (4) Excretory system. (5) Regeneration system. (6) System of synthesis. (7) System for sorting and transporting substances. (8) System for maintaining integrity. (9) Support-locomotion system. (10) System for protection from environmental factors. (11) System for receiving and transforming information. (12) System of genetic control. (13) Reproductive system. (14) System of development.

Some other functional systems that serve to carry out different tasks at various moments of cellular life may also be singled out conventionally.

It is more correct to call these systems structure-functional systems, rather than just functional systems. It is obvious that the whole complex of functions is accomplished through the interaction of highly-ordered structures and parts of cells. But we could still call them functional systems, insofar as this emphasizes the fact that various combinations of the main cellular parts can perform or participate in the performance of many kinds of different functions.

As an example, consider a system for sorting and transporting substances, which carries out the global function of distributing synthesized and incoming substances and transporting them to points of application. Let us trace the path of synthesized proteins in a cell. Each newly synthesized protein follows a specific path, determined by the signal segment of the polypeptide chain. The sorting of proteins begins with initial segregation, when a protein either remains in the cytosol or is transported into the other compartment. Those proteins that get into the endoplasmic reticulum (ER) undergo further processing and sorting. The treated proteins are packed into membrane vesicles at the terminal areas of the ER. The vesicles separate from the reticulum and are purposefully transported to their destination points—the Golgi apparatus, lysosomes, plasma membrane, or other. In the Golgi apparatus, proteins undergo additional processing and are packed into membrane secretory vesicles. After interflowing with the plasma membrane, these membrane secretory vesicles release a substance that contains the proteins. The vesicular transport system is used to transport proteins. The vesicles separate from one compartment, migrate, and then target-interflow with the other one, which has special receptors. Vesicles target-migrate in a cell by molecular routes, formed by protein fibers of the cytoskeleton.

If the newly synthesized proteins do not have signal sites for transport in organelles, they remain in the cytosol. Such proteins have signal sites that determine their place of localization. They are transferred to certain areas of the cytosol, undergo modifications, and are fixed by proteins of the cytoskeleton in an orderly way, forming enzymatic chains of metabolic fates. Fatty acids are covalently joined to some proteins, and the lipoprotein thereby formed may become part of the cell membrane structure. Proteins that have signal peptides for nuclear import are actively transported through nuclear pores, which recognize these proteins. A fairly complicated mechanism for sorting and transporting proteins into mitochondria is also based on the presence of a special signal in a polypeptide chain and the presence of the corresponding receptor on the surface of the mitochondria. The process itself is multistage and includes various molecular mechanisms. Old and defective proteins are immediately destroyed by the cell. Any 'destabilizing' amino acid on the N-end promotes adjunction of ubiquitin. Then protease combines with the marked protein and rapidly disintegrates it to amino acids. It is thus clear that the finely tuned system for accurately distributing protein molecules to the points of application is performed in cells. In cells, other systems sort and transfer many other substances used to maintain the strict order of cellular structures and processes.

Almost nothing in the cell possesses only one function. The number of performed functions is much higher than the number of organelles. In cells, against the background of basic functions, a reversible self-organization of various structure-functional blocks into a certain functional system may occur when necessary. If the situation changes, then cellular parts are easily reorganized to carry out other tasks. Thus, cellular parts, organelles, molecules, and their complexes are able to effect many variants of organization in dynamic systems to accomplish numerous complex functions. The molecular processes and interactions in all functional cell systems are extremely dynamic and possess a purposeful and interrelated character.

# Chapter 19
# Physiological Mechanisms

## 19.1 Physiological Processes and Mechanisms

A function is a controlled specific activity of a certain biological system, aiming to maintain and preserve the organism it forms a part of. For example, the heart is a complicated ordered system of interacting cells with various cellular differentiations. Its main function is periodic regular contraction and relaxation to ensure uninterrupted circulation of the blood in a multicellular organism.

Cells are the typical organizational units of life in all organisms. They are multiheterogeneous systems of interacting molecules and their complexes. Metabolism, biosynthesis of proteins and other essential molecules, and the realization of hereditary information and reproduction are only possible at the cellular level. To begin with, colonies of similarly structured cells were formed in the process of evolution, and then groups of cells gained specialization, depending on peculiarities of the environment, to become tissues of multicellular organisms. Cells possess various common and specific functions depending on differentiation. For example, metabolic functions (synthesis of macromolecules, transformation of energy, ability to divide, etc.) are common to all cells. Specific functions are typical, for example, for cells of different tissues. In particular, muscle cells contract, neural cells generate and transmit electric pulses, epithelial cells form various functional layers and surfaces, and cells of connective tissues ensure various functions, synthesizing specific molecules for the formation of intercellular substances. Thus, the aggregation of macromolecules in cellular systems endows cells with qualitatively new properties that are not inherent in the separate macromolecules and their complexes. It is a qualitative evolutionary step which has conditioned the emergence of a qualitatively new phenomenon—function, as a conceptually new feature of biosystems.

Tissues are systems of cells and intercellular substance, united by commonality of structure, function, and nature. Hundreds of different cell types make up the body of a multicellular organism. Four types of animal tissues are distinguished, depending on location, structure, and function: epithelial, neural, connective, and muscular. In this case we can also see that uniting various cells into dense groups where they may interact conditions the emergence of conceptually new properties and functions of biosystems. A quantity transforms into a quality. A quality transforms into properties. Properties transform into functions.

Organs are systems that consist of cells of different tissues joined together in order to perform certain functions. They are strictly ordered complexes of cells and tissues of various types, established during the process of evolutionary development for the purposes of carrying out specific functions. Organs are highly differentiated body parts located in certain areas and possessing certain functions. They are formed in the process of development from cells of various tissues. Higher animals have many organs, diverse in size and structure, and possessing numerous functions. For example, an eye is a sight organ, an ear is a hearing organ, a heart is a pump for the blood system, etc. Many organs perform several functions. In particular, liver cells produce proteins of blood, bile, and urine, as well as neutralizing toxic substances and many more. An organ consists of structure-functional units, which are the separate cells or integrity of ordered cells capable of carrying out the main function(s) of the organ. For example, the functional unit of the liver is the liver acinus, of the lung, it is the alveoli, of the kidney, the nephron, and so on. In the development process, organs are united into systems in order to be able to perform more complicated functions.

A physiological system is a hereditary fixed complex of cells, organs, and tissues, associated with common functions. In particular, the cardiovascular system consists of numerous variously differentiated cells that form a heart, various vessels, and capillaries. But the main task of this system is to ensure the motion of the inner fluids of an organism, which provides integration, regulation, and metabolism. Mammals have a number of physiological systems: neural, endocrine, immune, digestive, cardiovascular, excretory, etc.

A functional system is a dynamic integrity of physiological systems, various organs, and tissues that act together to achieve adaptive results useful for the organism. That is, functional systems are interacting structure-functional blocks which jointly maintain optimal homeostatic characteristics to assure adaptation, survival, and reproduction. Depending on the requirements of the organism, functional systems can be formed by various components of physiological systems. For example, for restoration of blood pressure after blood loss, the cardiovascular, neural, endocrine, excretory, and digestive systems work in concord. The interaction of structure-functional blocks conditions the high reliability of multicellular organisms.

The described characteristics of the main biological systems of multicellular organisms represent interrelation of structure and function, the logic of formation of functioning systems, and the emergence of certain functions.

The essence of this is the consolidation and interaction of various structural units of biological systems. The sum of the units possesses completely new properties. On this basis, functions emerge. But they would never become homeostatic without the expediency of their actions for an organism. Only the necessary functions become fixed in genetic material and become inherited.

Moreover, each function is strictly controlled by the organism via the neural, endocrine, immune, and other systems.

The sequence and interrelation of several functions, directed toward the achievement of a certain result, comprise a physiological process. For example, the process of breathing is achieved by regulated consecutive functions of the nasopharynx, trachea, lungs, alveoli, blood, and erythrocytes, and the tissue respiration of cells of various organs. A common goal of this package of functions is the transformation of energy.

A multicellular body is a complex system of interdependently functioning cells, tissues, organs, and systems of organs. It is the organism that is the unit and carrier of life. All the levels of its organization, i.e., molecular, cellular, organ, and others, work together and are coordinated by the neural, endocrine, and immune systems for the purpose of survival of the individual organism. Thus, an organism is a heterogenic supersystem that consists of subsystems of molecules, organelles, cells, tissues, organs, and body parts. An organism is structured on a hierarchical basis, that is, the simpler components make up the more complicated ones determining their qualitatively new properties. On the basis of this concept, the elements of the neural or endocrine system can regulate molecular and cellular processes and functions.

We may conclude from the above that, as a new property of living material, function emerges only at a certain stage of its development. Separate molecules do not possess functions. Neither molecules of proteins, nor nucleic acids, nor any other organic or nonorganic molecules possess any ability to maintain themselves, their environment, or a surrounding system. However, they do possess various physicochemical properties. For example, proteins are amphoteric electrolytes with a very stable primary structure and buffer properties. The majority of proteins are fairly soluble in water. Aqueous solutions of proteins are very stable homogeneous colloidal solutions in equilibrium. They are characterized by a low rate of diffusion, inability to pass through biological membranes, and high osmotic activity. They have high viscosity. Fibrillar proteins are inclined to form gels. Proteins are very stable in physiological conditions and able to regulate reversible changes of their own conformation. But none of this relates to functions—they are just properties. However, on the basis of these properties, and interactions with each other and with other molecules, proteins perform various functions in the organism, viz., structural, catalytic, contractive, regulatory, protective, transporting, and many others.

Therefore, properties 'become' functions under certain conditions. For example, the structural function of proteins becomes manifested once proteins have been united into organized complexes and systems on the basis of physical and chemical interactions. In particular, biological membranes are formed in this way. These are ordered complexes of lipid and protein molecules that have a certain qualitative and quantitative composition, united on the basis of polar and hydrophobic interactions. Separate molecules of this complex do not yet possess functions. It is only their organized totality and interaction that ensure various directed processes. Functions emerge on this basis. Thus, membranes may perform the following functions: barrier, transport, energy, reception, communication, etc. Therefore, it is the ordered assembly and interaction of the elements of biological systems that conditions the performance of specific functions. Furthermore, certain conditions are required for the manifestation of membrane functions, viz., an aqueous medium with neutral pH and moderate temperature and pressure. For example, catalytic functions of proteins occur only in an aqueous medium, at mild temperatures, and in the presence of substrates.

Thus, organisms are complex ordered systems that consist of several levels of organization. The main parts of living bodies are molecules, cells, tissues, and organs. Chemical, physical, and biological properties of the structural units and their combinations in specific environmental conditions (aqueous medium, pH, temperature, etc.) determine their various functions.

It should be noted that functions and processes are the derivatives of structure. Clearly, a certain material structure must first emerge, and it is only after that, depending on the conditions, that a function may manifest itself. For example, the function of muscle contraction requires the presence of the acto-myosin structural complex. That is, peculiarities of functions are completely determined by the properties of structure. The structure conditions the size, shape, aggregative state, and reactivity of systems, a stationary and potential characteristic of the living body. Function is a dynamic characteristic that reflects directed changes in material systems. Functioning is directed and regulated by biological, chemical, or physical processes that change the stationary characteristics of the system in time. The correspondence and interrelation between structures and functions are typical for all levels of organization in living organisms. This can only be understood and explained through the laws of development and evolution. However, the development process just answers the question of how complicated structures of tissues and organs can emerge and form, but does not answer the question of what conditions the emergence of functions.

Functions of non-living systems (excluding machines created by humans) are neither directed nor operated by themselves, while the functions and processes of living systems are precisely directed by the elements of these systems, and are directed primarily at maintaining the stationary condition. It is obvious that the basic functions of living organisms, such as nourishment, respiration, excretion, and others, serve to maintain integrity and homeostasis. Therefore, a precise correspondence between structures and functions is observed at various levels of organization (from organism to molecule). Moreover, every structure is optimal for performance of a specific function. A distinct correlation is observed between the level and complexity of organization of biological structures and the presence of corresponding functions.

Both single-celled and multicellular organisms face the same problems of survival during their lives: the need to grow and develop, maintain integrity and autonomy, reproduce, feed, breathe, etc. This conditions the presence of analogous structures, which may actually differ in structure, but perform the same functions. In particular, single cells have peculiar organs of motion (flagella, cilia, pseudopodia), organs of digestion (digestive vacuoles), organs of excretion, organs of self-regulation, protective organs, etc.

Thus, at a certain stage in the development of biological systems, on the basis of the physicochemical and biological properties of the elements of these systems, there emerges a specific controlled physiological activity, directed toward preservation, maintenance, and reproduction.

## 19.2 Mechanism and Processes of Development

Development itself is the mechanism of formation of very complex organisms through special processes of embryogenesis and post-embryogenesis.

Multicellular organisms are very complicated in terms of organization. For example, a human being consists of trillions of cells of more than 200 types, arranged in an orderly way. A human being has complicated organs that consist of millions of different cells, each of which is located in the required place and possesses the required function. It is hard to imagine how such a complicated organism could have emerged suddenly, or even how such a complex organ as a brain or an eye could have emerged. But everything becomes more or less clear once we realize that such complicated systems emerge only through a process of gradual step by step development on the basis of consecutive differential gene expression of the zygote. They do not arise as whole formations, but are formed from a small group of cells in an embryo through the processes of embryonic and post-embryonic development.

Gradually, gene after gene, cell after cell, process after process, tissues, organs, and the whole organism emerge and then become more complicated. At the beginning, after the first few weeks, a miniature human, consisting of thousands of cells and weighing only a few grams, develops from one cell. It is very tiny, but already has all tissues and organs, all body parts and extremities. This fantastic period of embryogenesis is determinative—a highly organized organism is virtually completely formed from one cell by means of consecutive divisions and differentiation. It then grows and by the moment of birth it is completely ready for autonomous existence. After birth it needs 16–18 years more to reach the limits of growth and development. It is apparent that the complicated purposeful process of development is the mechanism of formation of an integral organism with highly complex organization.

There is a wide variety of different types of individual development in Nature. However, all of them include different versions of the processes stated below, which lead to the gradual complication of the developing organism over several stages (see Table 19.1).

We shall now briefly define the main processes of development of multicellular organisms.

*Germinal period.* Gametogenesis is a complex of processes that ensures the formation of mature germ cells, a precondition for the reiteration of a cycle of

**Table 19.1** Main periods, stages, and processes of individual development

| Periods | Stages | Processes |
|---------|--------|-----------|
| Germinal period | Primordial germ cell | |
| | ↓ | ← Gametogenesis |
| | Gametes (ovules and spermatozoa) | |
| | ↓ | ← Fertilization |
| Antenatal period | Zygote | |
| | ↓ | ← Cleavage |
| | Blastula (single-layer embryo) | |
| | ↓ | ← Gastrulation |
| | Gastrula (multi-layer embryo) | |
| | ↓ | ← Hysto- and organogenesis |
| | Differentiated embryo | |
| | ↓ | ← Growth and development |
| | Fetus | |
| | ↓ | ← Birth |
| Postnatal period | Newborn | ← Growth and development |
| | ↓ | |
| | Juvenal organism | ← Pubescence |
| | ↓ | |
| | Pubertal organism | ← Ageing |
| | ↓ | |
| | Old organism | ← Death |
| | ↓ | |
| | End of vital processes, disintegration of highly ordered organism into simple molecules | |

individual development of future generations of organisms. Once the organism has reached sexual maturity, gametes may escape from the gonads and lead an independent form of life.

*Fertilization* is a process of fusion of haploid gametes that results in the formation of a diploid cell, or zygote, the initial stage of development of a new organism. Fertilization leads to two very important processes: activation of the ovum (impulse for development) and caryogamy, the unification of the father's and the mother's genomes. Fertilization conditions the beginning of new life. The mature organism, producer of gametes, eventually gets older and dies.

*Antenatal period. Cleavage* of the zygote is a series of rapidly occurring divisions. As a result, a large volume of cytoplasm and genetic material of the zygote divides into numerous smaller cells called *blastomeres*, which are similar or slightly different in size. They usually form a spherical structure or *blastula* (single-layer multicellular embryo).

*Gastrulation*. This is a complex of processes of division, growth, differentiation, and directed migration of blastomeres, which change their structure and location in relation to each other. This results in the formation of a gastrula that consists of three cellular layers (germinal layers): the outer layer is called the ectoderm and the inner layer the endoderm. Then, in the case of three-layer animals, the third layer is formed. This is an intermediate germinal layer called the mesoderm. All three layers subsequently originate the anlage of organs and tissues.

*Hysto- and organogenesis.* A complex of processes of division, interaction, and migration of the cells of the germinal layers, which gradually gain precise orderliness and form tissues and organs of an embryo. Many organs are formed from cells that originate from the different germinal layers. A differentiated embryo is formed as a result of organogenesis.

*Growth and development.* This involves the increasing size and mass of an organism by means of constant controlled division of cells and the gain in mass of intracellular substances, along with a gradual complication of organization and functioning. Growth and development occur at the molecular, subcellular, tissular, organ, and organismal levels. All levels of growth and development are under genetic control and ensured by differential gene expression during ontogenesis.

Thus, multicellular organisms are formed from the fertilized ovule, initially as small groups of indistinct and similar cells, which later transform into a large, differentiated, perfect organism as a result of growth and development. The following stages and processes tend to complete rather than continue the process of development of an individual.

*Birth* is the escape of an organism from the maternal organism.

*Postnatal period.* This is *growth, development, and pubescence,* wherein the organism achieves a certain size and functional capabilities and enters the reproductive period. *Ageing* is a process of gradual wear and degradation of the structure and functions of cells and the organism as a whole. *Death* marks the end of the processes of vital functions.

Thus, the formation and emergence of any living multicellular organism is only possible through the process of development. This extremely complicated process ensures a complex of many molecular, cytological, and physiological mechanisms of interaction, which unerringly direct the flows of matter and energy under the control of the genetic programs of the genome.

However, the genome does not contain certain information concerning development, for example, with regard to rates, quantities, sizes, forms, and places of localization. So where does this information comes from? It is known that, in the processes of development of a living body, progressive structure-functional modifications occur all the time, leading to an accumulation of order and the emergence of new information. This information, manifested through a specially organized structure, conditions a new stage of the development process. So any process of development is self-informing. Every preceding state carries information about the direction of future events. That is, apparently, the process of development is not based purely on genetic information. Genetic blueprints are the main elements managing a process, which thereafter, by means of changing body forms, develops rather as a 'snowball', on the basis of additional structurally conditioned information. But what limits the borders and defines the forms of information spread?

The mechanisms and processes of phenotypic life are described in this part. The transformation of matter and energy in living bodies is necessary for their development, functioning, and maintenance of integrity. The mechanisms of genotypic life will be described in the next part.

## Recommended Literature

1. Prigozhin, I.R., Stengers I.: Order out of Chaos. Man's new dialogue with nature. Moscow (1986)
2. Prigozhin, I.R.: The philosophy of instability//Problems of Philosophy. No. 6, (1991)
3. Trincher, K.S.: Biology and Information. Elements of Biological Thermodynamics. Nauka, Moscow (1965)
4. Bauer, E.S.: Theoretical Biology. ed. VIEM (1935)
5. Wiener, N.: Cybernetics and Society. Nauka, Moscow (1967)
6. Alberts, B., Bray, D. et al.: Molecular biology of the cell: Garland Science, New York (1994)
7. Lehninger, A.L., Nelson, D.L., Cox, M.M: Principles of Biochemistry, 2nd edn. Worth, New York (1993)
8. Hess, B.; Markus, M.: Order and chaos in biochemistry. Trends Biochem. Sci. **12**, 45–48 (1987)
9. Youvan, D.C., Marrs, B.L.: Molecular mechanisms of photosynthesis. Sci. Am. **256**(6),42–49 (1987)
10. Dressler, D., Potter, H.: Discovering enzymes. Scientific American Library, New York (1991)
11. Harold, F.M.: The vital force: a study of bioenergetics. W.H. Freeman, New York, (1986)
12. Mitchison, J.M.: The biology of the cell cycle. Cambridge University Press, Cambridge (1971)
13. Hyams, J.S., Brinkely, B.R. (eds.): Mitosis: Molecules and Mechanisms. Academic, San Diego (1989)
14. Zhabotinsky A.M.: Content fluctuation. Nauka, Moscow (1974)

# Part IV
# Mechanisms of the Invisible World of Information

# Chapter 20
# Biological Information and Cybernetics

## 20.1 Bioinformatics

*A. Thesaurus.* The notion of 'information' is a fundamental category, but it is hard to define. Information is usually defined as a set of data that can be generated, transferred, accumulated, perceived, or used. Information by itself is not material, but it is a property of matter, like its discreteness and movement, for example. Any object, phenomenon, or event that may cause a variety of interactions and states of many different elements can be a carrier of information.

Information can exist and be spread by different forms of material carriers: atoms, molecules, objects, fields, waves, vacillations, flows of particles, etc. It may also exist in the form of ideas, thoughts, fantasies, images, etc. That is, it can be a product of matter, but have no material nature. We can say that information is different manifestations of fluctuations of the surrounding space. Any changes of any dimensions are a source of information. However, information can manifest itself only in the presence of an object that can perceive it.

An endless source of information for living bodies is the total variety of their organized and non-organized dynamic surroundings. In addition, there are even more powerful flows of internal information in living systems. In the biosphere, several main flows of information can be noted (Sect. 2.8): (1) flows of external information; (2) flows of intracellular information: (a) flows of genetic information, (b) flows of molecular information, (c) flows of information about orderliness; (3) flows of intercellular information; (4) flows of information between organisms: (a) between specimens of one species, (b) between specimens of different species; and (5) flows of genealogical hereditary information.

In the 1970s, characteristic theoretical principles of semantic information were developed for biosystems. The basis of this theory was not only the idea of a code and channel of transfer, but also the properties of a receiver that perceives information, as well as assessment of its meaning (or semantic significance). This means that semantic information, which is perceived by a given living system, may be evaluated solely in terms of the system's own previously accumulated

G. Zhegunov, *The Dual Nature of Life*, The Frontiers Collection,
DOI: 10.1007/978-3-642-30394-4_20, © Springer-Verlag Berlin Heidelberg 2012

information. In other words, in order to adequately perceive information from external sources, a biosystem must itself possess some minimum knowledge (or minimum information). This minimum knowledge of a biosystem is referred to as its thesaurus. The thesaurus is an informational characteristic of the organization of a biosystem.

The initial thesaurus is a necessary condition for mastering and accumulating external and internal data for biosystems. In the process of information perception, systems react through a change in their state, and implement potential properties and functions. Information perception also contributes to the development and improvement of such smart systems. In this respect, the effectiveness of information exchange is determined by the properties of the information receiver, rather than by those of its source. Biosystems of any level of organization are sources of diversified signals that carry information about their organization and functions. However, these signals can be perceived as meaningful information only by those systems that understand it, i.e., they possess a specific thesaurus. Therefore, among living organisms, information is transferred on an "everything for everybody" basis and is perceived on a "for whom it may concern", basis, i.e., for systems that are capable of perceiving it.

Information appears (better to say, manifests itself) only with the emergence of objects or systems that possess special receptors for information carriers, as well as a specific thesaurus. The appearance of living systems led to the emergence of semantic information that is purposefully used by the thesaurus for self-preservation and survival. Accumulation of information and improvement of the thesaurus conditioned the progressive development of living bodies and their evolution. Peculiarities of external environmental factors (i.e., peculiarities of the quality and quantity of external data) led to the specificity of thesaurus formation in living bodies that live under certain conditions, and this defined the direction of their development. This was one of the reasons for the appearance of a variety of species.

*B. Cybernetic systems.* The science of control, connection, and processing of information is called cybernetics. It studies properties of various control systems without regard for their material basis. Properties concerned with controlling and being controlled are characteristic of biological objects, various societies, and various technical systems.

The main subject of study in cybernetics is information. Any phenomenon or event can be a source of semantic information, and it may often serve as a signal for some action. For example, molecules of a hormone are perceived as a signal by cells that possess specific receptors. This leads to activation of specific processes (like ATP synthesis) in cells, in response to the received command.

A cybernetic system is an organized and ordered set of interrelated and interacting elements of a system, able to perceive, remember, and process information, as well as to exchange it. Examples of such systems could be living objects (cells (Fig. 20.2), organisms, communities of organisms) or various devices, machines, computers, etc. Systems differ in their complexity and level of organization. The complexity of a system depends on the number of elements, their setup, or a variety of internal connections between them. Most complex cybernetic systems created by humans are known down to the smallest detail (computers, robots, conveyers, or tracking devices); yet, living bodies created by Nature have remained unclear until now, due to an enormous number of diversified system elements, manifold bonds between them, and the complex system of hierarchical organization. In many cases, the same part may be composed of several blocks or systems. For example, the most complex multicellular organisms are controlled by a brain, which is an independent complex system that consists of billions of no less complex cells. The functional unit of a brain is the neuron, an extremely complicated, independently functioning cybernetic system with an enormous number of different connections and very complex internal molecular organization.

Biological systems are probabilistic, because variants of their behaviour are difficult to determine due to the impossibility of making exact assumptions about the interactions and reactions of the manifold components of such systems, affected as they are by an enormous number of simultaneous physical and chemical factors.

Cybernetic systems are self-contained if their component elements exchange signals only between each other. Open systems like living organisms also exchange information with the external environment. For this purpose, animals have a complex system of analysers: visual, auditory, tactile, vestibular, etc. Every analyser is an intricate system which includes certain main elements such as receptors and an analysing centre. Receptors are special cells, specially designed nerve endings, or modified nerve cells located in various parts of a body. They are generally able to perceive only specific types of irritants. For example, modified nerve cells of an eye retina are sensitive only to electromagnetic radiation of specific wavelengths. Regardless of its exact nature, information perceived by receptors is converted into an electric current (translated into the universal language of living bodies), which is transferred to afferent neurons via nerve processes (conductive section of an analyser). This analysing centre is the main link of an analyser system. This is where incoming information is processed and converted into specific electric signals, which are transferred to an efferent neuron. This cell also generates an appropriate electrical signal, in accord with the received information. This signal is transferred via nerve tracts to effector organs, which react to the received signal through a specific action. For example, this may be a hormone secretion, contraction, excretion, enhancement, or inhibition of various processes or functions. The set of events in response to specific information received by a system is called a reflex.

Complex cybernetic systems, including cells and organisms, are capable of preserving and accumulating information that can be used afterwards. This property is called memory. Memorizing can be achieved in two ways: (a) through a change of state of the system elements, (b) through the change of its structure. In particular, the order of DNA molecular structure contains a large amount of information about the development, structure, and operation of various cells and multicellular organisms. In the process of expression (change of state of system elements), information is implemented in specific traits that determine the characteristics of a system. The brains of animals especially those of human beings, can accumulate and store large amounts of information that can be used for survival, operation, and reproduction. In addition, through memory, humans and humankind as a whole are capable of learning, accumulating a mine of information using artificial carriers, to store it for a long period, and to transfer it to future generations.

The transfer of information in the form of signals is carried out through connection channels. A connection channel is a medium used to transfer signals. For nerve regulation of muscle contraction, the signal is an electrical nerve impulse and the medium is the axon membrane. Various factors of the material world can be physical carriers of signals: molecules, electromagnetic radiation, mechanical movement, electrical impulses, gravity, radiation, etc. In information transfer and processing, one form of signal can be transformed into another. For example, the energy of photons of visible light in the eye retina is converted into electrical impulses, which are transformed into messenger molecules of the nervous system in the nerve endings of brain neurons. These act on the membrane of a specific neuron, generating an electric current that circulates in a specific part of the cerebral cortex where it causes specific visual images. Information should not be distorted during perception, transfer, and processing, so that it may provide an adequate reaction of the system. This phenomenon of the correspondence between a signal and the reaction is called isomorphism. Problems with isomorphism cause inadequate reactions of the system.

Depending on the peculiarities of the system reaction, signals can be informative (report information) or executive (convey a command to action). In particular, with vision, we mostly receive informative signals, while molecules of hormones, for example, give commands for action by target organs.

Signals of various kinds will provide specific informational action only if special receptors are present. For example, there are special receptors for hormones, but they are present only at target cells. The rest of the cells thus remain insensitive to the given signal. External signals are also received only with the help of appropriate organs or special cells. Thus, for example, eyes are receptors for a specific range of electromagnetic waves; ears for oscillations of specific frequency, etc. An enormous amount of information is not received by living organisms, due to the lack of special receptors. In particular, mammals are insensitive to gravity, magnetic fields, flows of neutrinos, various microwave frequencies, X-rays, and other types of radioactive emanations. (However, some of the above-mentioned agents do have a direct impact on components of living bodies).

Humans can be called the most perfect "antennae" in terms of the perception of various signals. Through various sense organs and various receptors, they can detect a large amount of different kinds of information. The smallest quanta of information of a different kind is intensified several times and turned into the energy that triggers motion in cells, systems, parts of bodies, and the whole organism. Besides this, humans have steadily increased antenna capabilities with the instruments and methods of research. With the help of an electron microscope, visual acuity is increased a million times, allowing us to see the contents of cells as easily as we would admire a landscape.

Humans have even learned to perceive information that is unnatural to them. For example, information received with the help of X-rays can be used in experimental physics, chemistry, and biology, for diagnostics in medicine, etc. Ultrasound serves to obtain an abundance of valuable information through special devices that radiate and receive sound waves, thereby demonstrating a wide range of processes. In addition, humans are able to use the following as sources of information: heat radiation, gravity, and flows of neutrons and neutrinos, or other sources that are not inherent to them. Even though people do not possess the appropriate analysers for these flows, they have learned to transform physical sources of information into images that are easily perceived by standard human sense organs. For example, X-rays leave visual traces on special film and the molecular composition of blood can be visualized with the help of electrophoresis or chromatography.

Any message consists of a combination of simple signals. A complete set of such signals is called an alphabet, whereas a separate signal is called a letter. In particular, four nitrogen bases of DNA (adenine, guanine, thymine, and cytosine) constitute an alphabet for coding genetic information, and specific bases are letters of a genetic code (A, G, T, and C). A sequence of three bases (codon) of a structural gene encodes one specific amino acid. For example, mRNA codon AUG encodes methionine. A succession of codons encodes the primary structure of a specific protein. The process of encoding involves recording information with the help of an alphabet. The translation of this information into another alphabet is called recoding, and the complete deciphering of communication is called decoding. For example, translation (formation of a specific polypeptide based on an RNA matrix) is recoding. The encoding phenomenon enables the use of a small alphabet for storage and transfer of large volumes of information. As an example, in computer technology, a digital language consists of just two symbols (0 and 1, binary code), and the volume of applicable information is tremendous. Encoding, transfer, and decoding of most biological information are based on a binary code (yes, no). Receptors either perceive signals or not, transform them or not, and transfer them or not. A signal is either analysed or not, a system reaction either exists or it does not. Electroexcitable tissues and all membrane mechanisms of cells work solely on the principle of a binary code. "Yes" stands for the presence of an electrochemical potential in a membrane, and "no" stands for its absence.

Perception, transformation, and transfer of information are connected with energy consumption. Failures in energy exchange by neurons can cause pathologies of the nervous system and many other functions due to mistakes in information processing and transfer.

*C. Control.* Control is essential for the purposeful modification of a cybernetic system. Control is a specific informational impact that initiates a standard programmed reaction of a system, resulting in achievement of the required purpose. The essence of programmed control is the initiation of movement and interaction of significant amounts of matter, as well as a transfer and transformation of large amounts of energy, controlled and directed by a small amount of matter or energy that carries the information. Informationally controlled processes are inherent to any cybernetic system. This can be the transfer of hereditary traits in biological systems, leadership in a team, control of machines, conveyers, etc. The principles of control are similar for subject systems of varied nature and complexity—from regulation of molecular processes in cells and various functions in organisms to the working principles of computers and spaceships. Any control system consists of a controlling organ, a subject of control, and a channel of connection between them. A controlling organ processes information and produces a command which is transferred through the appropriate channels to the controlled object. Connection is made with the help of physical and chemical processes that carry information. After receiving a signal, the system moves into the required state. For example, a human brain receives information from thousands of external and internal receptors every second. It processes it consciously or unconsciously and a command is then generated and sent out in the form of electric nerve impulses via axons to controlled objects, such as the muscles of the hands or feet, endocrine organs, or other body parts. As a result, an organism proceeds to a qualitatively different state and completes an action in order to achieve a certain aim.

Control of cybernetic systems can be carried out without the interference of humans, but on the basis of a specific program specified by them. In living organisms, many processes can be regulated unconsciously. This option is called automatic control. In living organisms, thousands of different biochemical and physiological parameters are automatically controlled and regulated (e.g., the amount of water in cells and tissues, ionic and cellular composition of blood, frequency of heart contractions, blood pressure, electric potential of cells, amount of ATPs formed, digestion, formation of urine, and many others). This leads to constant maintenance of the internal environment in living cybernetic systems, otherwise known as homeostasis, which is a necessary condition for the stable operation of a system, even under changing conditions. The most widespread and efficient systems for maintenance of homeostasis are closed control systems with feedback, when a controlling organ processes information received from the outside or from other components of the system, and also receives signals from the controlled object through feedback. Feedback is transfer of information from the controlled object to a controlling organ. Positive and negative feedback are distinguished. With positive feedback, the return signal enhances the communication process that causes a transition of the system to a new level, or an avalanche process, while negative feedback slows and stabilizes the process, hindering its development.

Systems with negative feedback are capable of self-regulation. Living organisms are closed cybernetic systems with feedback, i.e., they can control themselves on this basis. This can be found on all levels of organization of living bodies. For example, on the molecular level, when a specific concentration is achieved, product molecules of enzymatic processes inhibit the work of enzymes that catalyse this process. Besides the substrate centre, such enzymes have a specialized centre of allosteric regulation for the products of biochemical processes.

Inside cells, the concentration of all molecules is maintained at a constant optimal level, which is also regulated through the feedback principle. The amount of synthesized hormones in cells of endocrine organs is controlled by the action of these molecules contained in blood. The constant body temperature of mammals is regulated on several levels, through feedback, special receptors, and diversified effector organs.

   *D. Information Value of Matter.* Any law of Nature is a set of information about the properties of specific material bodies, systems, or processes that are understood by human beings. Each body contains a specific amount of information. Elementary particles, atoms, molecules, or other elements of systems interact naturally, and when bonds are formed and the system has moved into a different state, they transform and accumulate new information. In contrast, scientists study bodies and their properties, discover information contained in them, and establish various laws.

   The same concerns the dynamics of physical and chemical transformations of substances. Any transformation is related not only to transformations of matter and energy, but also to changes of information. For example, in biochemical reactions of protein synthesis in cells, not only do material DNA—RNA—protein transformations take place, but information encoded in genes is modified, transferred to a different carrier, and then used for the formation of polypeptides.

   Information cannot exist without a material carrier. Any information, set of knowledge, reflex, or idea is connected with matter and material processes. The medium of information is always a material body or a field. Information is enclosed in these like potential energy, and like potential energy, it may be manifested or not. Information manifests itself through various fluctuations of the natural environment, in the form of waves, oscillations, body movement, molecules, atoms, etc.

   As a part of Nature, information has many different levels of organization: the level of the Universe, galaxies, stellar systems, planets, bodies on a planet, molecules, atoms, and elementary particles. The world around us contains a tremendous amount of information, part of which is already understood by humans. Living organisms on the planet Earth are "a drop in the ocean" compared with all the matter in the Universe. But this is a qualitatively specific state of matter. Living bodies possess a number of unique properties and mechanisms. They have more levels of order and organization, i.e., they possess a considerably larger amount of accumulated information. They include all information from the lower levels of organization, typical of non-living bodies, connected with elementary particles, atoms, and molecules. The next level of organization of living bodies is the formation of macromolecules: DNA and RNA. This is an essentially new level of organization of matter, midway between non-living and living. This is the highest level of "informatization" of matter. The indicated polymer molecules possess aperiodic (Schrodinger, 1955) orderliness in the structure and alternation of monomers, i.e., they possess a language, or a specific arrangement of nucleotides, the sequence of which determines the recording and accumulation of an enormous selection of information. This information is sufficient to create many millions of species on the basis of hundreds of thousands of different proteins. This is already the next level of "informatization" of matter. Proteins contain not only linear information concerning the alternation of amino acids, but also spatial information. This is the information regarding their surface (peculiarities of size, shape, charge, etc.), which predetermines the recognition and natural interaction of various molecules, as well as internal information (regulatory centers, catalytic centers, functional groups of atoms, mobile segments, etc.) which determine a countless multitude of functions and properties of proteins. Supramolecular complexes, such as biological membranes, ribosomes, enzyme complexes, and organelles, are built on the basis of this complex of information and synergy mechanisms. These structures can be considered as the next level of informatization in living matter. Organelles are constructed on the basis of all the previously listed information, and are of course its accumulators. Cells are highly ordered heterogeneous open systems. They consist of naturally arranged atoms, molecules, macromolecules, and organelles. These systems are formed on the basis of the aforementioned information complex, controlled by the genome. They contain all the listed information as a standard of specially modified material space.

*Carriers and information.* As a consequence, organisms are living carriers of information. From this perspective it is interesting to note some analogies between the properties of artificial media, such as compact discs (CD) containing computer files, and natural carriers–living bodies that contain genetic information:

1. There is no information without a carrier. *There is no life without a body..*
2. The value of information can be much higher than the value of the carrier itself. *The value of the permanent genome is much higher than the value of a temporary phenome carrier.*
3. Information from a CD can be copied repeatedly or rewritten to other media. *Genetic information is replicated many times and is transmitted to the next generation of carriers.*
4. Information from a CD can be read selectively. *Genome expression occurs differentially.*
5. Information may or may not be displayed. *Genetic information may or may not be expressed.*
6. Special tools and mechanisms are required to extract information from a CD. *Gene expression requires special enzymes and cytogenetic mechanisms.*
7. Energy is required to extract information from a CD. *Gene expression requires energy.*
8. The slightest change in the CD structure leads to a modification of the playback. *Mutations in genes cause changes in the phenotype.*
9. Information from a CD "materializes" and is visualized on the screen of a monitor or in the form of sound. *Genetic information is materialized in the form of elements of the phenotype.*

These analogies indicate the dual nature of living bodies, the indivisibility of information and matter, the greater value of information in relation to the carrier, the similarity in the processes operating the information, and the potential eternity of nonmaterial information in comparison to the vulnerable material carrier.

However, there are also fundamental differences between non-living and living information carriers. Artificial carriers cannot reproduce. Such carriers have no exchange of matter and energy with the environment. Non-living carriers cannot restore their structure, and cannot be subject to natural selection and evolution.

*E. Information and level of organization.* A cell is a system of interacting molecules. This implies extraordinary dynamics, not only of matter and energy, but also of information. The required molecules are synthesized and transformed on the basis of their genetic information. Only the necessary molecules enter or leave the cell, by virtue of the "information awareness" of the membrane. Molecules interact with each other through their linear and volumetric information. Enzymes, created by DNA molecules, condition strict selectivity and order of molecular processes, e.g., synthesis of specific proteins, breakdown of glucose, oxidation of tricarboxylic acids, etc. All the above processes involve the use and transformation of different information flows. From this point of view, cell metabolism can be viewed as strictly directed, interrelated interacting flows, not only of matter and energy, but also of information. Multicellular organisms include all the above-mentioned background information and, in addition to this, an enormous amount of information that is inherent only to them. This is due to the natural laws of structure and organization of different complex living bodies, as well as the presence of various organs, tissues, and body parts, and thousands of different cells. Millions of species of living organisms possess thousands of different functions. All this encompasses a mine of information. Thus, the level of organization of living bodies is determined by both the quality and the quantity of information used for their construction. In particular, primitive unicellular organisms and bacteria have the smallest genomes, whereas multicellular organisms have larger genomes (although there is no direct correlation).

The degree of development and complexity of organisms is also determined by the quality and quantity of information available to perception, analysis, use, and generation. Animals, especially highly organized ones, are able to perceive and analyse large amounts of information. For this purpose they have different sense organs (smell, touch, vision, hearing, etc.). Some animals have organs for perceiving ultrasound, invisible electromagnetic radiation, or electric fields. Some are able to sense and orientate themselves in magnetic fields or sense radioactive radiation. Plants are photosensitive. They can perceive temperature shifts and possess seasonality. All the aforementioned factors serve as specific signals for living organisms. They carry a tremendous amount of information. Perceiving that information with the help of special receptors and analysers, a living body can perfectly orientate itself in space and time, avoiding unfavourable factors, choosing the best time for reproduction and growth, finding food and sexual partners, etc. All this supports the main aim of life—survival. Most living organisms are able to specially generate and distribute various signals (information) in their vicinity. Thus, they can contact and interact with each other even at great distances. This is very important for survival, guaranteeing the reproduction and maintenance of genetic continuity. In addition, all living organisms have massive internal flows of information.

In other words, we may say that the ability to create, perceive, accumulate, process, and use information is one of the most important features of life. It is one of the factors which distinguishes the living from the non-living. Non-living subjects do not feel the above-mentioned flows of information; they do not have similar organs that would perceive and process information. They have no internal flows of information. In principle, they perceive some information, or better to say, some impact, like radiation or heat, but in a passive manner. Non-living bodies are not capable of analysing such information and responding with strictly standard and appropriate reactions. That is why non-living bodies are not able to avoid unfavourable factors or adapt. Like closed systems, they do not exchange matter, energy, or information with the external environment. The levels of organization of non-living bodies are considerably lower, implying a considerably smaller amount of information in their structures and the inability to manipulate it. Only certain machines and devices created by humans possess any such ability, and then only to a certain degree (e.g., machines with program control and computers). However, they are not autonomous. Such bodies are entirely dependent on humans and incapable of operating information flows independently.

*F. Information and entropy.* From the point of view of thermodynamics, a developed living organism is an open system, whose negentropy (order) grows with time, depending on the growth of used and accumulated information. This conditions the possibility of improving and complicating the structure, as well as implementing more and more complex work processes. The ability to use information for reduction of entropy is also one of the distinctive properties of living bodies.

When an organism approaches the adult state, other genetic programs become activated. They are directed against spontaneous processes of disintegration, that is, against increase of entropy. Such information and processes based on it maintain the integrity of an organism for a certain time. Termination and abnormalities in performance of these programs result in an irreversible increase in entropy and imminent deterioration of the living body. Finally, everything ends with death: complete destruction of an organism and the triumph of entropy over matter and information.

Unlike mortal living bodies, the immortal life phenomenon remains in a constant process of development. This process tends to constantly complicate, accumulate information, and lower the level of entropy in the Integrated Life System (Fig. 2.15). Therefore, throughout the process of evolution, the informational content of the life phenomenon, which is characterized by structural order, increases with time. In this sense, there is a general tendency from chaos to order, from simple to complex, from entropy to negentropy. Thus, information and entropy may change in any direction in the process of ontogenesis within individual carriers of the genome. However, in the Global System of Life, information tends to steadily accumulate and entropy to decrease.

**Fig. 20.1** "Bioinformational big bang"—the basis for the development and spread of life. Its mechanism is a super-high-speed chain reaction of replication of DNA molecules, when a living body appears "from nothing"

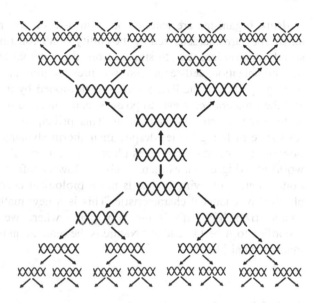

Thus, living bodies are like clots of information, in fact, of materialized genetic information. They are built on the basis of information. They live and survive on the basis of information. In living organisms, every level of organization corresponds to a flow of information. The molecular level is the principal such level, mainly concerning DNA molecules. An "informational big bang" of a set of these molecules from a specific genome in a specific informational field of the environment causes a cascade of consecutive transformations of matter, energy, and information leading to the creation of a certain living body (Fig. 20.1). From this point of view, living bodies are products of the interaction of genetic programs with information from the environment.

*G. Genetic informational programs.* The notion of a genetic program stands for a specific plan or scheme for a distinct succession of elementary actions recorded in specific genes of a given genome. Genetic programs are implemented with the help of special molecular mechanisms of transcription, translation, etc., which form a strictly conditioned process, gradually unfolding in time and leading to the creation of certain structures and functions (e.g., programs of growth and development of a body, programs of structural–functional organization of specific organs, behaviour programs, and others).

Genetic programs are very rigid. They constitute the rules of absolute behaviour of biological systems. Figuratively speaking, this is a code or constitution of behaviour for all constituent biosystems. Molecules are moving and transforming in metabolic chains and cycles according to strictly genetically conditioned protein paths. Deviations are practically impossible. Cells divide, grow, and interact on the basis of genetically determined mechanisms. Organisms grow, function, and reproduce under the strict control of genetic programs and genetically conditioned mechanisms of the nervous, endocrine, and immune systems.

Thus, organisms are autonomous, multi-level, self-regulating cybernetic systems. They are characterized by the ability of all constituent elements to react in a strictly appropriate way to signals from the external and internal environments. Self-organization, self-regulation, self-preservation, and self-reproduction are the main properties of the living beings, conditioned by the ability of living systems and their material elements to possess, generate, and transfer information, as well as to perceive, process, and use it. This principally new quality conditions the resistance of living bodies despite their thermodynamic imbalance. The constant flow of matter and energy takes place in a non-organic environment, while in the world of living creatures there is also a flow of information. The super-informational content of living bodies is not a biological characteristic, and not even a physical or chemical characteristic. This is a new mathematical category, which characterizes life! This is the very place where we have to search because, according to Albert Einstein; "Nature is the realization of the simplest conceivable mathematical ideas".

## 20.2  Living Computers

A countless number of machines created by human intellect and hands are currently operating around us: cars, airplanes, TV sets, mobile phones, computers, etc. Machines are autonomous devices built from structural and functional units that carry out well defined and relevant actions and work. Living bodies are also autonomous, built from their own units (organelles, cells, and organs), and possess the ability to fulfil predicted actions. They have specific standards of composition, structure, and operation and possess definite types of behaviour, standard functions, and common principles for carrying out work. Everything we have said so far allows us to consider organisms as specific, constantly working living machines.

Cells are the smallest organisms, and units of multicellular bodies. Generally, they are considered to be complex mechanisms for directed controlled transformation of substances and energy. We may consider cells through the analogy with computers, as systems that also transform information. Computers are machines used as universal devices for accumulating, processing, transferring, and applying various types of information.

Cells of various kingdoms of living organisms are built of largely similar units and fulfil a set of standard predictable actions. These minute machines are molecular systems, because they are built from standard sets of molecules and their complexes. The work of cell-machines is mainly focused on themselves, i.e., on support of their own metabolism and homeostasis, as well as on their interaction with the environment. The specificity, interconnection, and purposefulness of the work of all the components of cells is ensured by genetic programs of original sets of DNA. Since cells have a pan-controlling genome, they conform to the notion of program-controlled machines.

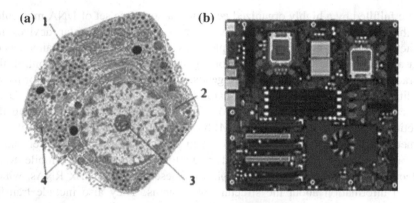

**Fig. 20.2** Living computer. **a.** The cell. *1*—The superficial apparatus of a cell contains a numerous receptors, which assimilate semantic information from the external environment. Selection, distribution, cascade enhancement, and analysis of information take place here. Furthermore, information is channelled to executive units and applied. *2*—The protoplasm is a colloidal matrix that integrates all system elements. It is a unit for communication and execution (Fig. 3.3). Information is transformed and exchanged here. It is realized in activation, deactivation, or regulation of all functional units. *3*—The nucleic apparatus of cells is a storage unit for genetic information and its extraction, operation, and transformation. This is a system unit which contains a genome processor and DNA memory. It is also a control unit for all the functions of cells through special genetic programs. The controlled objects are all the elements of the system. Management operates through self-contained feedback systems. *4*—Mitochondria, membranes, and special enzyme systems form the energy supply system for the processes of transformation of matter and information. **b.** Part of a motherboard of computer. This is also an open system, consisting of interconnected units, analogous by functions to protoplasm

The genome of a cell contains recorded information about the composition, organization, and operation of the entire system. Throughout the process of cell life, information is unpacked, deciphered, and transferred from DNA to proteins, where it is implemented in the form of specific traits and properties. Many molecules of proteins and enzymes, as well as their complexes, are involved in the processing of genetic information. Special macromolecules transfer information from DNA to RNA, and then to proteins. In other words, special molecules operate with genetic information! Thus, one might say that living cells are molecular-operational machines (Fig. 20.2).

It is still difficult to imagine in full detail how such cellular computers are organized or work. However, we may offer a number of analogies and suggest some hypotheses.

A computer system unit contains a multitude of details, the main ones being a processor and an information storage system (memory). The performance capabilities of computers depend on the running speed of the processor, as well as on the volume of operational and long-term memory. A computer processor is a controlling device, and constitutes one of the main parts of an operating system. The main function of processors is automatic management of computer operation with the help of programs located in the memory. The role of a cellular processor can be fulfilled by a highly organized genome—a complete set of DNA molecules together with enzymes and regulatory proteins (special molecular devices for extracting information). The very complex structure of a genome contains tens of thousands of genes, as well as various genetic programs. These control the interaction and selective expression of genes, depending on development or living conditions. Long-term memory (or hard drive) corresponds to a set of DNA molecules in which an enormous amount of information is recorded using the genetic code. Operative memory (RAM) has a distinct analogy with a complex of formed RNA. Information is extracted from DNA located in the nuclei and is accurately transferred, without any damage to the structure, to mobile RNA molecules which then act in the cytoplasm. These include matrix RNAs, which contain information about the structure of proteins. They also include transfer RNAs that carry information for targeted binding of specific amino acids (protein monomers) and supply ribosomes with amino acids. And they include ribosomal RNAs containing information about the organization and functioning of ribosomes. Various different kinds of information can be "downloaded" from the DNA "hard drive" to the "operative memory" of RNA. This results in the formation of a broad spectrum of proteins that carry out a multitude of functions.

All the data and programs on a computer are recorded in the form of files or sets of files. Files are areas of information on carriers, e.g., magnetic or optical discs. All the files in machine memories have unique names. The analogues of files in biosystems are genes. These are specific segments of DNA which carry particular information. Every gene also has a unique name and localization. Various genetic programs are recorded and stored by sets of genes (genetic networks). By analogy with archive files, genes and genetic programs can also be temporarily inactivated and archived in the form of mitotic chromosomes or in the form of a gamete genome. Under appropriate circumstances, the archive is unpacked and starts to function.

*Computer software* is a succession of commands and data, which are perceived and processed by a computer, and are intended to achieve specific results. Programs determine definite functions of machines: from a simple text editor to the most complex control programs of a spacecraft. *Genetic programs (software) of living computational machines* comprise a qualitative and quantitative set of genes (files) in chromosomes (hard discs). This is a controlled succession and regularity of expression of specific groups of genomes. Realization of the above mechanisms can bring about a countless multitude of response options. In every case of expression, diverse groups of RNA and proteins will be formed, and this conditions the appearance of a mass of different cells with the potential to carry out a multitude of functions. *A set of programs* for a specific type of computer determines the full diversity of their application, just as a set and combinations of genetic programs determine the diversity of living organisms, their traits, and properties. In other words, frogs have their own set of programs, and cats also have their own, and this conditions their morphofunctional individuality.

*The operating system* is the main program (a set of basic programs) that controls the work of a computer as a whole. For example, IBM PCs mainly use

*Windows* operating systems. The role of a *cellular operating system* can be ful-filled by a principal set of genetic programs located in the genomes of all organisms. These are programs like *replication, transcription, and translation* (let us refer to this as the *RTT operating system*), which determine the structure and strategic properties of living bodies: metabolism, homeostasis, and reproduction. The multitude of other programs is auxiliary (e.g., synthesis of phospholipids, glucose oxidation, production of ATP, etc.).

*The material basis of a computer* is the set of all parts, units, junctions, and so on, made of special substances and arranged in a certain manner in a limited space. *The material basis of a cellular computer (hardware)* is its highly organized internal molecular contents. First and foremost, it is a system of structural and functional proteins. In the aqueous environment, proteins form a highly organized and ordered *colloidal solution* (gel) that possesses the properties of a liquid crystal. In various parts of a cell (nucleus, mitochondria, cytoplasm, etc.), it possesses different protein compositions and different physical and chemical properties, and fulfils various functions. In the micro-space of the cytoplasm, all composite ele-ments are constantly moving around, there are billions of molecular interactions, and at the same time there are transformations of matter, energy, and information. In general, such an *aqueous protein matrix* is the material basis for the cell's operating system.

*Computer viruses* are special self-propagating programs. Such programs can damage or destroy programs and files stored in the computer memory. Cellular viruses consist of DNA or RNA and are active only in a nucleus of cells infected by them. They can also be considered as "self-propagating programs" which use the material basis and operating systems of cells for their own self-reproduction. They also often cause mutations in the genes (damage or destroy files) of host cells.

A minimal unit of computer information is called a *bit*. A bit has a value equal to 0 or 1. Any information may be encoded by a succession of zeroes and ones. Computer technology uses a binary code. It is amazing that such an enormous amount of information can be recorded just with the help of two symbols! The *memory unit* in modern computers is a byte. Bytes contain 8-bit binary numbers—00000000, 00000001, ..., 11111111. One byte is recorded with 8 binary units of information in the form of zeroes and ones.

A symbol of one of the four nitrogen bases of nucleotides, with the help of which DNA molecules are built, can be considered as a *unit of genetic information*. These are A, T, G, and C (adenine, thymine, guanine, and cytosine). This means that biological systems have a quaternary code for encoding information. In other words, living systems can theoretically record and store a considerably larger amount of information in comparison to a binary system of coding.

*The unit of memory of a cellular computer is a triplet*—stable combinations of three nitrogen bases, for example, ACC, TGA, GAC, AAA, CAT, etc. Every triplet (byte) encodes a specific amino acid, and a succession of triplets in a DNA segment encodes a specific polypeptide chain. Polypeptide chains form a basis for the structure of thousands of proteins which independently and in different combinations determine the countless multitude of traits and properties of living organisms. Thus, it is evident that biological computers possess an original language which is more powerful than a computer's.

A *computer display* is a device that allows visualization of information on an electronic screen. For living computers, the "displays" are their *bodies*. Genetic information is visualized in the form of the phenotype—the aggregate of internal and external characteristics of a given specimen. For example, for a cell, these are the size, shape, number of chromosomes, peculiarities of feeding, presence of biochemical and biophysical processes, capacity for movement, presence or absence of a specific function, and many more. All these properties are peculiar reflections of concrete genetic information.

The e*nergy supply* of computers is provided from the outside by power from an external circuit or a built-in battery. Like any machine, cells also use energy for all kinds of activities. *The energy supply system of cells* is an enzyme system for controlled catabolism of nutrients and photosynthesis in plants.

The *external data input device* of a computer is represented by a keyboard, which is analogous to the superficial cellular apparatus. This apparatus contains numerous receptor-keys, which transmit special signals when impacted, thereby conditioning the corresponding processes and functions.

All computer units are connected together by hundreds of *electrical communication channels*. The circulation of electric currents along various communication channels, controlled by programs, maintains the coordinated work of all computer systems. Intracellular units and systems of macromolecules also function together and are well coordinated by virtue of *intracellular communication channels*. For example, these are strands and tubules of cytoskeleton that permeate all internal cell space. Various molecules move along their surface in a highly ordered and controlled manner, e.g., substrates for the group of enzymes of a specific metabolic path. But the main role of communication channels is played by *the system of water channels*. As thick as several $H_2O$ molecules, such channels form spontaneously around various macromolecules, around strands of cytoskeleton, and in a space within the cytoplasm structured by proteins. Here, several water layers are formed, creating a continuous dense network which connects the internal contents into an organic whole. Depending on the orientation of the water molecules in a specific layer, only the molecules of certain substances can be selectively "dissolved", and then transferred quickly and directly to the required area with a speed comparable to the speed of diffusion in pure water. Flows of molecules along such channels transfer and transform not only matter and energy, but also information. All the above-mentioned channels are highly dynamic and, depending on the needs of cell, they can disintegrate or they can be renewed, redirected, or created anew. This provides the dynamics and lability of matter, energy, and information transfer, thereby making the cell highly sensitive and mobile. We may say that the structured aqueous-protein matrix is a *global communication system*, which unites all molecules into an integrated functional system.

The analogies presented here confirm that cells are amazing cybernetic micro-devices that can operate extremely well, not only with matter and energy, but also with information. In other words, we may say that they are living computers, capable of copying themselves, developing independently, and evolving.

## 20.3 Info-Genetic Mechanisms and Processes

Info-genetic processes are simultaneous, parallel, interrelated transformations of genetic material and information. These processes unite and guide all genomic and phenomic elements of the Integrated System of Life. In this section, we shall stop for a moment to examine their purpose in life.

*a. Processes intended for recreation of living bodies.* Self-reproduction is one of the main properties of living bodies. The central molecular mechanism of reproduction in all types of living organisms and cells is *replication*. This is a complex, multi-stage process of doubling of DNA molecules that takes place in the nuclei of cells during preparation for their division. Dozens of different enzymes participate in this process. Most of these enzymes are standard for all living organisms. Each double spiral DNA in each chromosome of the karyotype unwinds and forms two separate strands of polynucleotides. Each strand then acts as a matrix for exact synthesis of a complementary strand. As a result, two identical DNA molecules are formed. Doubling of genetic material forms the basis and condition for cell division.

The process of even division and distribution of chromosomes in daughter cells is carried out by mitosis. This complex process comprises several stages and is described in detail in Sect. 18.1. It results in the appearance of two identical genomes of daughter cells in the place of one maternal cell. A new organism is then gradually created through progressive division of cells and differential expression of the genome. Thus, directed transformations of genetic information determine its continuity, as well as the continuity of its material carriers—living bodies.

*b. Processes directed at development.* Development provides for the formation of a large complex organism from a zygote, containing billions of diversified cells and dozens of organs and body parts. This cumulative result is conditioned by many complex cytogenetic processes (e.g., cleavage, gastrulation, and others), which are described in Sect. 19.2

Therefore, multicellular organisms are formed from a fertilized egg, initially as small indistinct groups of homogeneous cells, which then gradually transform into a large differentiated organism as a result of info-genetic mechanisms.

*c. Processes directed at the maintenance of metabolism and integrity.* The main condition for the existence of living bodies is a constant exchange of substances and energy. Billions of different chemical and physical transformations of molecules take place each second inside every cell. A multitude of organic substances is purposefully destroyed for extraction of energy. Other substances are synthesized to ensure constant renewal of the composition of cells and the organism. Special genetically controlled mechanisms act to support long-term existence and normal functioning. The main executors of homeostasis and integrity maintenance are proteins. They are encoded by DNA genes and special processes exist for their expression.

The main molecular-genetic processes, aimed at ensuring metabolism, functions, and the maintenance of integrity in living systems, are described in Sect. 21.1. The above-mentioned info-genetic mechanisms are determinative in ensuring the autonomous existence of living organisms.

   d. *Processes aimed at evolution.* Life is a process aimed at the future. Despite the mortality of individual representatives of species, their genome continues to exist in different bodies for millions of years. Life as a phenomenon, as a process of development of the Global Genome and Phenome, has already existed for 4 billion years and will continue to do so in the future. Life has evolved simultaneously with our planet as an integral part. For this purpose, living beings possess molecular-genetic mechanisms (recombination, mutation, hybridization, transgenesis, infobiogenesis, and others) which allow them to continually adapt, survive, generate information, exchange information, and evolve under the circumstances of constant variability of the material world. These mechanisms of evolution have already been described in Sects. 23.2, 23.3, 24.1

   Thus, it is evident that the different info-genetic processes have a common molecular basis and principles of realization. These very processes are the ones that unite and direct all structures, mechanisms, and processes in the Integrated Life System.

# Chapter 21
# Genetic Information

## 21.1 Attributes of Genetic Information

Living organisms are able to create and maintain a high level of organization, grow, differentiate, and propagate thanks to genetic information, which is stored and reproduced, and transferred to future generations through the molecules of DNA. Any orderliness of matter in biological systems appears on the basis of energy, as well as information concerning qualitative composition, quantity, and arrangement of system components. Genetic information is information about the structure and functions of living organisms encoded in DNA and inherited by every generation from their ancestors in the form of the totality of the genome.

Discrete units of genetic information are genes. Chromosomal DNA contains thousands of genes that carry information about all the proteins synthesized in cells. Sets of genes of a specific organism form a genotype. This information is encoded in DNA in the form of particular successions of nitrogen bases, called a genetic code. The idea that information is stored in DNA, implemented through its transfer to an mRNA and then to a protein, is considered the main doctrine of molecular biology. Thus, the basic law of molecular biology is an informational law. Genetic information, recorded in a mathematically precise linear sequence of DNA nucleotides, is rewritten into a linear sequence of RNA nucleotides, which is later translated into a linear sequence of amino acids in polypeptides, and from there into a three-dimensional structure of proteins. The proteins formed in this way determine biochemical, physiological, morphological, and many other traits of organisms that make up their phenotype.

The flow of genetic information and its relationship with *material processes* can be represented as follows:

$$DNA \xrightarrow{Replication} DNA \xrightarrow{Transcription} RNA \xrightarrow{Translation} Polypeptide \xrightarrow{Folding} Protein \xrightarrow{Expression} Trait$$

G. Zhegunov, *The Dual Nature of Life*, The Frontiers Collection,
DOI: 10.1007/978-3-642-30394-4_21, © Springer-Verlag Berlin Heidelberg 2012

- *Replication* is a process for copying genetic information. It is a transfer of information within one class of nucleic acids, involving *synthesis of identical molecules of DNA for the purpose of transfer to offspring* (*italics show material processes*). Replication is the molecular basis for all types of reproduction.
- Transcription is a process for rewriting genetic information from areas of DNA molecules to RNA molecules. It is a transfer of information between different classes of nucleic acids. *It is a process for the synthesis of complementary strands of RNA from specific DNA genes.*
- Translation is a process for transfer of genetic information from mRNA to polypeptides. It is an information transfer from one class of molecules to another. *It is a synthesis of linear polypeptide molecules from amino acids.*
- Folding is a process for transformation of linear two-dimensional information into a spatial three-dimensional form. *It is the totality of molecular processes that lead to the formation of tertiary and quaternary protein structures.*
- Expression is a process exploiting the linear and 3D information of protein molecules for interaction with other molecules. *Interactions and associations between different molecules condition the manifestations of traits.*

It is important to mention that the path from a gene to a protein is more or less well understood. The same cannot be said about the details of transformation of a complex of synthesized proteins into a functioning macrostructure.

*Genetic information possesses unique properties and characteristics:*

1. *Linearity in the recording and reading of information.* An enormous amount of information is written down in terms of linear successions of nucleotides in DNA molecules, and this is read strictly in one direction, from the $5'$ end to the $3'$ end, translated into a linear succession of RNA polynucleotides, and then translated into the linear succession of amino acids of a polypeptide.
2. *A large volume of information.* There are about 3.2 billion nucleotide sequences in the haploid genome of human DNA. This is sufficient for encoding approximately a million genes. According to the biogenetic law (embryogenesis is a condensed repeat of phylogenesis), animals store information about structure, processes, and functions of all phenotypes of their ancestors in their genomes!
3. *Compactness of packaging.* An enormous amount of information is recorded with the help of small nucleotide molecules in the DNA macromolecule, and this is sufficient for the development of a large complex multicellular organism with thousands of characteristics. For example, in a human being, this information is recorded in 46 molecules of DNA (46 chromosomes), which are freely placed in the microscopic nucleus of a cell that can only be seen with a microscope.
4. *Amount of "irrelevant" information.* From a tremendous encoded set of information, only approximately 5 % is used in the expression of certain genes.
5. *A large amount of duplicated information.* The genotype of most living organisms is diploid. In other words, it consists of two genomes: maternal and

paternal. Every gene is represented by two copies. That is why, when one mutates, an organism still preserves viability, because the majority of mutations are recessive. Moreover, many important genes are represented in several copies for higher stability (for example, genes of histones or certain peptide hormones). Some genes form repeated tandems and clusters.

6. *Highest stability of information.* The above-mentioned peculiarities of genotype organization guarantee high stability for the storage of genetic material, as well as protection from the influence of unfavourable factors. Furthermore, in the nuclei of cells, there are special enzymes for repairing damage to DNA molecules, and this also significantly increases the stability of information.

7. *Lability of information.* Genetic information possesses changeability, which may result in the appearance of new phenotypes. The reasons for the variability of genotypes are mutations, replication mistakes, and recombination of DNA, as well as the creation of new combinations of genes as a result of gametogenesis and fertilization.

8. *Accuracy of information transfer.* Many species of living organisms have existed on Earth for hundreds and thousands of millions of years, preserving all their structural and functional features. This means that genetic information is able to preserve accuracy and stability for such long periods of time. This phenomenon is ensured by periodical rewriting and editing of DNA during replication in the process of each reproduction cycle.

9. *Procreation accuracy.* Special enzymes control accuracy of replication, transcription, and translation. Many mistakes are already rectified during the listed processes. If there is still an "incorrect" protein, it is immediately recognized and destroyed by special proteases.

10. *Molecular-digital principle of recording.* From the point of view of cybernetics, biological systems use molecular-digital principles for recording and reproducing information. The recording matrix is a DNA molecule, and molecules of nitrogen bases play the role of digital symbols. In computer informatics, only two symbols are used (0, 1), and the volume of applicable information, for example, in the Internet, is simply tremendous. In biological informatics, four symbols are used (A, G, T, and C), whence the volume of recorded information may be billions of times more.

11. *The matrix principle for realizing information.* The main macromolecules are constantly synthesized by cells in large amounts. These are primarily proteins and nucleic acids. A matrix concept of synthesis is used to ensure rapidity and accuracy of these processes. In this case, one of the molecules, or part of it, serves as a template for mass production of the necessary molecules, one after the other. In this way, complementary strands of DNA are formed during replication (the matrix is each DNA strand), RNA molecules—in the process of transcription (the matrix is an area of one DNA strand), and polypeptide molecules—during translation (the matrix is an mRNA). In the above-mentioned process, there is not only matrix synthesis of specific molecules, but also a matrix principle for exact and quick transfer of information from one molecule to another.

12. *Catalytic mechanism of information realization.* The genetic information of DNA is implemented through basic processes: transcription, translation, and expression. Dozens of special molecular catalysts, or enzymes, serve the above-mentioned complex of processes for sequential information transfer.

13. *Consecutive principle of information realization.* In the process of realization of genetic information, products of previous stages condition the activation of subsequent stages. For example, RNA molecules needed for the synthesis of proteins are formed by transcription. These proteins, in turn, are necessary for the formation of enzymes, which are needed to ensure metabolism and form cells, which are in turn needed for the formation of organs and to ensure their functions, and so on.

14. *Multiple augmentation of information in the process of realization.* Insignificant changes in DNA molecules during the process of mutation, such as the replacement of one nucleotide, cause the erroneous inclusion of an amino acid into a polypeptide. A spatially "incorrect" protein molecule thus forms. This protein does not function or functions poorly. Therefore, one of the metabolic processes is damaged, followed by modifications to some cell function, then an organ, and in the long run, changes to some macro-trait, its disappearance, or the appearance of a new trait in the organism.

15. *Universality of genetic information.* The language and the methods of recording, storage, and realization of genetic information are the same for all living organisms, from bacteria to humans.

16. *Continuity in the transfer of genetic information in living organisms.* Termination or damage to the flow of genetic information cause the death of organisms, populations, and species.

17. *Integrity of genetic information.* A set of chromosomes, genes, and intergene areas of all DNA molecules form a genome that functions as a comprehensive whole.

18. *Genome information determines phenome characteristics.*

19. *The totality of genomes* of all living bodies forms an integrated system of the Global Genome, which is an integrated system for circulation of genetic information. Circulation is implemented by reproduction, as well as the horizontal transfer of genes.

It is thus evident that genetic information is the paramount natural phenomenon. The condition for its existence is molecular processes that ensure permanent interconnected circulation of matter, energy, and information. The link between DNA and the unique enzymatic molecules of matrix processes is a prerequisite for the realization of information in the form of living manifestations. DNA molecules without key enzymes are dead, in the same way that it is impossible for biocatalysts to appear without the appropriate DNA genes. *So which came first— matter or information?*

## 21.2   Genetic Code

The uniqueness of the structure and functions of millions of different cells in living organisms is conditioned by the specificity of their set of proteins. Cells are able to synthesize individual sets of proteins, as well as specific proteins by means of a specific portion of genetic information recorded in their own DNA molecules. This information exists in the form of a linear and regular sequence of nitrogen bases of nucleotides in DNA strands, and is called a genetic code.

The genetic code of DNA has the following fundamental characteristics: (a) *Triplet*: three appropriate nitrogen bases, called a triplet or codon, encode one amino acid. Codons are located next to each other without interruption. This means that the position of every amino acid in a polypeptide strand is dependent on the position of a triplet in a DNA. At the same time, the genetic code is 'singular'. For certain amino acids, one triplet code can be substituted for another. This also ensures the stability of living organisms—many spontaneous abnormalities in the genetic 'text' have a high probability of not affecting its meaning and not causing dramatic consequences. (b) *Specificity:* every separate triplet encodes only one specific amino acid. (c) *Disjointness:* the nitrogen base of a specific triplet is not usually part of another one (in some microorganisms, the possibility of a "drift of reading frame" can be observed); (d) *Punctuation marks:* the genetic code has no "punctuation marks" between encoding triplets in structural genes. However, it has many different signal sequences between other functional areas of DNA. Three variants of triplets (ATT, ACT, and ATC) do not encode amino acids, but behave as specific stops for a process of information reading. (e) *Universality*: numerous studies confirmed the universality of the genetic code, despite small differences of DNA code in mitochondria and bacteria. In other words, all living organisms from viruses to plants and mammals use the same triplet to encode the same amino acid. This is one of the most convincing pieces of evidence for the unity and common-ality of origin of all animate nature. (f) *Redundancy:* the four nitrogen bases adenine (A), thymine (T), guanine (G) and cytosine (C) in combinations of three can form $4^3 = 64$ different codons in DNA molecules. Since there are 64 possible variants of codons, but only 20 amino acids are used, the same amino acid can be encoded by different triplets (synonym codons). (g) *Collinearity*: the succession of amino acids in a protein corresponds to the succession of triplets in a mature mRNA. (h) *Unidirectionality:* the process for reading the information of a genetic code from a matrix strand of DNA goes only in one direction—from the 5′ end to the 3′ end. (i) *Compactness:* a tremendous amount of information is recorded in a macromolecule of DNA with the help of small molecules of nucleotides. It is sufficient for the development of a huge multicellular organism. (j) *Molecular-digital principle of data recording*: the recording matrix is a DNA molecule, and molecules of nitrogen bases serve as digital characters. In biological informatics, four characters are used (A, G, T, and C).

Thus, the genetic code is the language of life (Fig. 21.1), which is used by numerous genomes for recording, storage, and duplication of tremendous amounts

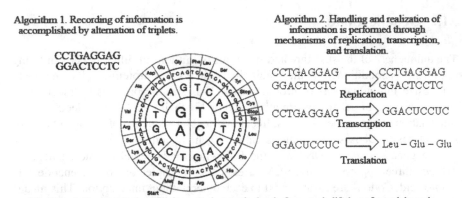

Algorithm 1. Recording of information is accomplished by alternation of triplets.

CCTGAGGAG
GGACTCCTC

Algorithm 2. Handling and realization of information is performed through mechanisms of replication, transcription, and translation.

CCTGAGGAG ⟶ CCTGAGGAG
GGACTCCTC ⟹ GGACTCCTC
Replication

CCTGAGGAG ⟹ GGACUCCUC
Transcription

GGACUCCUC ⟹ Leu – Glu – Glu
Translation

Algorithm 3. Communication of living organisms at the level of genotypic life is performed through mechanisms of gene transfer within the Global Genome (Fig. 4.8), due to universality of the language of life.

**Fig. 21.1** The alphabet and algorithms of the incredible language of life. This ensures, not only the recording (*1*) of information and its transformation (*2*), but also the molecular and genetic communication (*3*) between all living creatures inside the Global Genome. This language is used to write all the informational programs of the full diversity of life. Many more programs will be written with this language for gradual replacement of outdated organisms that are no longer compatible with the evolving reality

of biological information. Such information is the basis for repeated reproduction, maintaining order, and ensuring properties and functions of phenomes. That is, the Integrated Informational System of the Global Genome, the product of which is a Global Phenome, exists on the basis of the genetic code.

## 21.3  Genetic Material

All representatives of the living world have their programs of reproduction and development recorded in the complex arrangement of genetic material—a set of highly organized macromolecular structures based on nucleic acids.

The majority of cells of a multicellular organism (99 %) reside in an active state at stage $G_1$ of the interphase. At that period of the cell cycle, the genetic material of the nuclei is in the form of chromatin. Separate fibrils of chromatin present a complex of DNA and proteins in a 1:1 mass ratio. The number of chromatin strands corresponds to the number of diploid sets of chromosomes. Every strand is attached to the proteins of the nuclear membrane. Though DNA molecules are very long, they are packed by means of histone proteins into a superspiral of considerably shorter length. A specific area of the DNA winds around nucleosomes, which are formed by histone proteins. DNA strands continue from one nucleosome to another. Intertwining inside a nucleus, they form a chromatin network.

Approximately 1 % of cells in multicellular organisms are in the division stage. On average, division time lasts about 24 hours. When cells are preparing for division, chromatin spirals to form denser and more compact strands, called chromosomes. Chromosomes are rod-like bodies, which exist in this form only during mitosis. They may be viewed under a light microscope. Chromosomes carry genes that are units of hereditary information. If all the DNA of one nucleus in a human cell were laid out end to end in a straight line, it would be approximately 1.5 m long. After DNA replication, this quantity doubles! How is it possible to strictly order these 92 long and intricate strands and equally divide them between daughter cells? This happens by virtue of chromatin superspiralization mechanisms and very efficient packaging into small dense bodies. In this way, long intricate strands are packed into 46 small chromosomes.

Chromosome dimensions vary from species to species. In the metaphase, chromosomes of different organisms are from 0.5–33 μm long and from 0.2–2 μm thick. Generally, plant chromosomes are larger than animal chromosomes. Chromosomes in different pairs within the same cell may vary in size. The length of a chromosome depends on the molecular mass of DNA, and on the level of chromosome compaction.

The totality of metaphase chromosomes of a certain species or individual forms the *karyotype* (Fig. 21.2). Karyotypes of various organisms are rather conservative. Despite a sexual process, mutation, and recombination of genetic material, only the allele composition of DNA actually changes slightly. The chromosome and gene composition remain rather stable. That is why there have been no major changes in phenotypes throughout millions of generations. This has maintained the existence of various types of living organisms for millions of years. Only significant chromosome and genome mutations impact the phenotypes of their owners. Generally speaking, these are unfavourable mutations, the possessors of which do not survive or reproduce.

On the other hand, genetic material is very plastic. It is susceptible to various influences and regulations and manifests controlled genetic activity. In different cells of the same organism, different combinations of genes from the genome are expressed, different proteins and their counterparts are synthesized, and specific proteins appear. This provides the morphological and functional variety of hundreds of diversified cells in multicellular organisms.

Genetic material is very dynamic and demonstrates reversible changes of structure, form, and function during a cell cycle. *During $G_1$, interphase is represented by a network of active chromatin. *During the preparation of a cell for division (phase S), the chromatin network is doubled (DNA replication). *During the prophase, chromatin becomes dense through spiralization, and loses its activity. *By the end of prophase, chromosomes are formed from chromatin, constituted by separate, shorter, and thicker strands. *During metaphase, they become even shorter and more compact, obtaining an X-shape and a specific structure and size. *During anaphase, chromosomes divide into two chromatids. *During telophase, they untwist again and form an active chromatin network, which is typical for the interphase. The cycle can then be repeated many times.

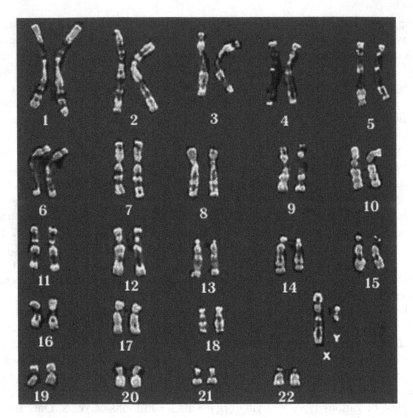

**Fig. 21.2** Human karyotype. The main elements of a genome—chromosomes. Mitotic chromosomes above are all a means of saving and transferring information. Genetic material is neatly organized, packaged, and preserved. Genetic information is duplicated and prepared for transmission

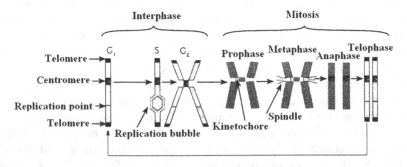

**Fig. 21.3** Diagram of chromosomal cycle. Transformation of genetic material of a genome during the cell life cycle. This process may also be referred to as a "genome cycle". It leads to the exact distribution of genetic information between daughter cells during division. Each daughter cell of a new generation receives one of the two copies of DNA. In other words, the processes of the chromosome cycle provide for the continuity of DNA in the neverending series of genome generations

Such repeated operations of genetic material are called *chromosome cycles* (Fig. 21.3).

As a result, there is an exact distribution of genetic information between daughter cells in the division process. Each new generation of daughter cells gains one of two copies of DNA molecules. In other words, the chromosome cycle can provide continuity of DNA in an endless series of cell generations.

The biological significance of chromosome formation and transformation is the formation of kindred molecules, i.e., replication, packaging, and equal distribution of hereditary material in the resulting cells. Chromosomes are just a convenient way to divide genetic material into two equal parts for the purpose of further transfer to daughter cells in order to form new phenotypes. This is a perfect mechanism for duplication of genetic information and its transmission to further generations of carriers. Indeed, it is an ideal method for reproduction of genomes and their phenotypic framework.

# Chapter 22
# Genes and Genomes

## 22.1 Genes

Genes are the units of genetic information, heredity, variability, and evolution. They directly or indirectly determine the development of all traits of an organism. Transfer of genes over a number of generations through the process of reproduction ensures propagation of organisms and transfer of parent traits.

According to the modern concept, genes are stable DNA segments that fulfil specific intragenome functions. For example, they encode RNA, enhance or weaken the action of other genes, participate in the shift of nucleotide sequences, etc. Before cell division, all genes are duplicated in the process of DNA replication, and then transferred to offspring. Genes consist of various functional segments, for example: promoter, operator, exons, introns, etc. Every gene occupies a well defined place (locus) in a specific chromosome. Different genes have a different qualitative and quantitative composition of nucleotides. Genes can exist in the form of several alleles that determine variants of traits.

Structural genes do not participate directly in the synthesis of proteins; they are matrices for the formation of messengers—RNA molecules that participate directly in the synthesis. During the transcription, genes are recoded into specific areas of RNA. All types of RNA participate in the synthesis of proteins, which provide the structure and metabolism of cells, and condition the development of traits based on the following scheme: gene → RNA → protein → trait. Monogenic traits are determined by one gene. For example, hair colour is determined by the presence and the intensity of expression of a gene that encodes a protein-type pigment called melanin. Polygenic traits are a manifestation of several genes. In particular, the dimensions of an animal body are dependent on many genes that condition the synthesis of various proteins. There is no specific gene for such complex traits as intellect, obesity, skin colour, etc., or a separate gene that would determine membership of a particular species or population.

A set of genes of a specific organism forms a genotype. Its differential expression in the process of development causes the formation of a phenotype—a

set of traits  and characteristics of a body. Information in genes is encoded by a
specific succession of triplets of nucleotides—a genetic code. The smallest genes
consist of several dozen triplets, for example, genes of tRNA . Genes of the larger
macromolecules rRNA  and mRNA  include several hundred and even thousands
of nucleotides.

The majority of genes in cells remain in an inactive state. Only a small portion
of genes (3–5 %) is active and may be transcribed at any given time. The quantity
and quality of gene functioning depends on the tissue the cell belongs to, the
period of the life cycle, and the stages of individual development. Manifestations
of gene activity are based on the presence of proteins in a cell or traits  of an
organism.

Structural genes (of thousands of different types) carry information about the
structure of specific polypeptides. The largest group is composed of genes that
encode enzymes—up to 32 % of all structural genes. Among such genes, there are
regulatory genes, whose products regulate the work of other genes. The action of a
given gene is strictly specific, i.e., a gene can encode only one amino acid
sequence and handles the synthesis of one specific polypeptide (not counting
alternative splicing). Some genes possess pleiotropy of actions, determining the
development of several traits  at the same time. The extent of gene action is
determined by the dependence of the intensity of trait manifestation (expression)
on the quantity of a specific allele (for example, in polyploidy).

The activity of a gene can be influenced by both external and internal factors.
Genes can recombine and mutate, providing changeability. However, DNA
molecules are subject to repair, which is why not all types of gene damage cause
mutation. In transgenesis, genes can be transferred, built in, exist, and express
themselves in "alien" genomes.

It is important to emphasize that the genotype, while being discrete (consisting
of separate genes), functions as an entity. Various genetic programs are recorded
and stored with the help of a set of genes. The units of expression of the genome
through different traits  are not the specific genes, but rather their systems, or
genetic networks.

Thus, the existence of genes conditions the most important processes for fixing
and operating genetic information—the method for managing the flow of living
matter.

## 22.2  Genetic Individuality

There is a wide variety of species of living organisms on earth. They exhibit
tremendous phenotypic differences, despite the common nature of life. There are
thousands and millions of species of single-celled organisms, worms, insects,
mollusks, fish, birds, or mammals. It is the individuality of proteins that underlies
this huge range of phenotypes. Different proteins form different structures, different structures condition different forms, and different enzymes and structures

condition different metabolic paths, properties, and functions. The variety of structures, properties, and functions conditions the form and the ecological niche of an individual's environment. However, all proteins of any given representative of a species, not to mention their quantity and quality, are controlled and handled by individual genomes—a full set of structural and functional segments of DNA, as well as programs and mechanisms of their expression that structure the phenotype of a concrete species. In particular, the genome of a human being contains approximately 35,000 genes, but there are no exclusive genes among them that would determine the development of a man alone. In other words, phenotypic and genotypic individuality are connected not only with genes, but also with some other factors, such as the mechanisms regulating expression. We may list several factors that condition genetic and phenotypic individuality on the basis of the peculiarities of the genome.

   A. *Genetic individuality of organisms of different* species *is conditioned by the following*:

1. Different number of DNA molecules.
2. Peculiarities of physical and structural characteristics of various DNA molecules and their complexes.
3. Specifics of nucleotide sequences in DNA, including both their qualitative and their quantitative composition.
4. Peculiarities of a karyotype.
5. Qualitative and quantitative composition of genes.
6. Qualitative and quantitative composition of alleles.
7. Peculiarities of intergenic segments of DNA.
8. Specifics of gene interaction.
9. Properties of functional networks of genes.
10. Peculiarities of the regulation of genes and genetic networks.

   The genetic individuality of different species is primarily conditioned by a karyotype—the chromosome complex of the given species, with all its peculiarities. The particular features of specific DNA molecules and chromosomes can vary significantly, depending on the nucleotide composition, length of a molecule, density of packing, and protein composition of the chromatin. It is quite possible that such peculiarities can possess as yet unknown manifestations concerning their impact on genetic expression. In particular, this can refer to a genome—the complete set of genetic material, all interacting DNA, and all genic and intergenic segments, which can have many unknown physical and chemical characteristics, each with its own biological effects.

   The number of genes in the genomes of organisms of different species is not the main characteristic or reason for phenotypic differences. For example, many plants, fish, rodents, and primates have a number of genes that differ only by a few percent. However, there are major phenotypic differences, displayed in hundreds of phenotypic traits. Genotypic differences between the closest primates (chimpanzees) and humans are as little as 1 %, but they result in great morphological, behavioral, and intellectual differences. Such essential differences are probably connected with the presence of specific genetic networks and peculiarities of their expression. An important role is also played by epigenetic systems, cytoplasmic heredity, protein heredity, structural heredity, and linked inheritance.

B. *Genetic individuality of organisms of one* species *is conditioned by:*

1. Variations in the allele composition of genes.
2. Single nucleotide polymorphism.
3. Specifics of intragene segments of DNA.
4. Peculiarities of gene interactions.

Even representatives of one species can have variable nucleotide sets in the DNA of their chromosomes. Nucleotide variation in genes conditions multiple allelism and polymorphism of traits. Variations of nucleotides in intragene segments and their effects have not yet been well studied, but there is evidence that even changes in nucleotide sequences in introns and "senseless" repeats may cause phenotypic effects.

The main type of genetic polymorphism within a species is single nucleotide polymorphism (SNP). These are versions of DNA in different individuals where only one pair of nucleotides has been changed. For example, such polymorphism is a primary source of differences between people. Differences in one pair of nucleotides can be found in any area of DNA: in exons, introns, intragenic areas, and repeats. The reason for their appearance is determined by mutation. In human populations, about 1.42 million SNPs have been identified and mapped. It has been established that people of the same sex are 99.9 % identical in nucleotide sequences. Thus, only a 0.1 % difference in nucleotide sequences conditions significant phenotypic variations in individuals. Differences in just one nitrogen base between certain segments of DNA underlie genetic diseases, sensitivity to or defense from certain agents, adaptive reactions, and hereditary predisposition to multifactorial diseases. Genetic individuality conditions protein individuality, which in turn affects the size and shape of living bodies, peculiarities of their structure, different functions, capabilities, metabolism, etc.

Genetic individuality (genetic homeostasis) of species and individuals is maintained over millions of generations due to special molecular and cellular mechanisms for protecting individuality. However, such protection is not always absolute. Foreign genes can be introduced into a genotype. The nature of multicellular organisms protects intensively from such disadvantageous possibilities, because otherwise it would be easy to lose genotypic and phenotypic individuality and disappear from earth. In order to avoid such events, evolution has resulted in the acquisition of special mechanisms that hinder transgenesis, mutation, and interaction of chromosomes. In particular, there is an entire complex system of molecular protection of DNA. This system includes repair enzymes, such as restriction enzymes, nucleases, DNAses, ligase, etc. These eliminate diverse mutations in DNA molecules. Furthermore, the ends of all chromosomes are protected by special end sequences called telomeres, which hinder the interaction and fusion of genetic material in chromosomes. The extremely complex immune system is also one of the most powerful methods of protection of multicellular organisms from foreign genes. Special cells and molecules of immune systems can recognize "genetic aggressors" and neutralize them, hindering possible transgenesis, parasitism, transplantation, and loss of genetic individuality. Whole groups of native, but genetically modified cells are often annihilated. Such mechanisms maintain genetic individuality for a long time, providing phenotypic individuality and stability over many generations.

There are also physiological mechanisms for supporting genetic individuality, for example, the impossibility of interbreeding between representatives of different species, or the impossibility of complementary combination in a zygote of haploid sets of chromosomes belonging to different species.

Thus, all organisms, populations, and species have inherent genetic (informational) individuality, maintained by special mechanisms at several organizational levels over long periods of time. Genetic individuality conditions the specificity of a set of proteins, whose properties and functions directly or indirectly ensure phenotypic individuality.

## 22.3  Expression of Genes

Gene expression is a set of cytogenic processes and mechanisms for implementing information, as a result of which genes manifest their potential in specific phenotypic traits  (Fig. 22.1).

A gene is generally inactive, but when a particular protein is needed, a concrete gene becomes activated, which conditions the manufacture of this protein. Cells thus have a mechanism that controls the quantity of any protein at any appointed time. Synthesis of proteins is also regulated by internal and external factors.

The concept of an operon, as a regulated unit of expression, was developed experimentally and theoretically for prokaryotic cells by F. Jacob and J. Monod in 1960. An operon is a succession of special functional segments of DNA that encode and regulate the synthesis of a specific group of enzymes of one metabolic fate, for example, glycolysis. An operon consists of the following structural parts: a gene regulator, which controls the formation of a protein regulator; a promoter— an area of DNA which affiliates with RNA-polymerase and initiates transcription; an operator—a segment of the promoter that binds the protein regulator; structural genes—segments of DNA which encode the mRNA of specific proteins; a termination part—a segment of DNA that contains a signal for ending transcription.

Gene regulators are important elements of an operon. They condition the synthesis of regulatory proteins under the impact of cellular factors. Such proteins can promote or prevent the connection of RNA-polymerase to a promoter when binding with some nucleotide sequences of a DNA operator. When the protein regulator prevents an enzyme from joining onto a promoter, it is called a repressor. In this case, a negative control of transcription  takes place on behalf of a gene regulator. If the protein regulator assists the connection of RNA-polymerase to a

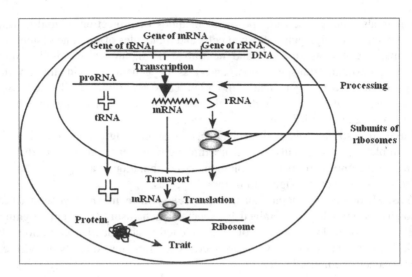

**Fig. 22.1** Simultaneous transformation of genetic material and biological information during the process of protein synthesis and formation of traits

promoter and the beginning of a transcription process, it is called a protein activator, and a positive control takes places on behalf of the gene regulator.

In addition, substances of non-protein nature (effectors) participate in the processes regulating gene expression. They interact with protein regulators and modify their ability to connect with an operon. For example, this could be an end-product of a metabolic process. Effectors are classified according to the results of such an influence either as inducers, which promote transcription, or as core-pressors, which prevent it.

*Peculiarities of gene expression in eukaryotes*. The principles of expression and its regulation are the same for both prokaryotes and eukaryotes. However, eukaryotes, especially multicellular ones, are very complex organisms, and the expression of their genes is considerably more complicated, differing significantly in certain details. In particular, the following peculiarities of expression can be noted in eukaryotes:

1. The eukaryote genome is considerably more complex. For example, the haploid genome of a human being consists of about 35,000 genes, located in 24 (22 autosomes + xx or xy) chromosomes. Prokaryotes have only one chromosome and just a few hundred genes.
2. In the cells of eukaryotes, the nuclear membrane divides the processes of transcription and translation spatially: chromosomes are located in the nucleus and ribosomes in the cytoplasm. Gene expression in eukaryotes includes many more stages, which is why they have a number of regulation mechanisms that are absent in prokaryotic cells, e.g., processing (see below).

3. The rate of expression of some genes is affected by gene amplification. This is a multiple increase in the number of copies of similar genes for the purpose of intensifying the synthesis of molecules needed at a certain time. For example, repeated sequences of DNA include hundreds of copies of rRNA and tRNA genes.

4. Structural genes of eukaryotes contain introns, continuous sequences of nitrogen bases which do not encode amino acids. Between them, there are sequences that do encode an amino acid—exons. RNA transcribed from a gene has both introns and exons. It is called a pro-mRNA. Its intron areas are removed by nucleases, and the areas that carry information (exons) are connected together. The complex of processes of pro-mRNA treatment and its transfer into an mRNA is known as processing.

5. The process of controlled cutting out of introns and connection of exons is called splicing. In different tissues and cells, various segments of a gene can be cut out and bonded together, which provides a way for several different mature RNA molecules to form. This mechanism is called alternative splicing. It allows several versions of polypeptide chains to be formed from the same gene.

6. In eukaryotes, there is no complete operon organization of genetic material. Genes of enzymes for a specific metabolic process can be located in various chromosomes. Generally speaking, they have no common regulation system in the form of a gene regulator, operator, and promoter. This is why mRNAs synthesized in eukaryotic nuclei are monocistronic. Regulation of gene activity in eukaryotes is more complicated, since several gene regulators participate in this process at the same time. In other words, the regulation of eukaryote transcription is combinatory. For example, the DNA molecule of a eukaryote has a special area near the promoter. This area includes about a hundred pairs of special nucleotides (promoter element). A special protein, the transcription factor, associates with this segment of a molecule and this provides for successful joining of RNA-polymerase II to the promoter.

7. The genetic systems of a eukaryotic organism, which provide the formation of any phenotypic trait, are genetic networks. Genetic networks are a set (system) of genes that control the appearance and manifestation of some property or characteristic of a living body. As a rule, the genetic network that controls a specific feature of an organism comprises several dozen to several hundred genes. Certain regulators of gene networks, e.g., encoding enzymes of a specific metabolic path, can activate transcription of the whole cassette of these genes. Regulation of genetic networks exploits a feedback principle. Any organism has a wide variety of genetic networks that control molecular, biochemical, physiological, and morphological features of the organism. And every genetic network has mechanisms that ensuring its regulation, based on changes in the functions of some groups of genes that are a part of this genetic network.

8. Another peculiarity of eukaryote genomes is the presence of special "amplifying" segments of DNA called enhancers. These can be located a long way

from the structural genes. In turn, both the pre-promoter element and enhancer are also regulated by corresponding regulatory proteins. Some protein regulators have a coordinating impact on the activity of many genes, i.e., they possess a pleiotropic effect.

9. The eukaryotic genome complies with regulatory impacts of the endocrine system of an organism. Many hormones are transcription inducers. In the first place, these are steroid hormones, which reversibly bind with protein receptors that carry them to a nucleus. Such a complex connects with a specific area of chromatin that is responsible for gene regulation. For example, the effect of testosterone activates genes that determine the development of an organism of the male type.

10. Another peculiarity of regulation of eukaryotic gene activity is associated with the formation of chromatin—a complex of DNA with proteins. In this form, as part of the nucleus, genes are incapable of transcription, so partial decompactization of chromatin and weakening of bonds with histone proteins is a necessary condition. However, the complete nucleosome organization of chromatin is not lost in the transcription process.

11. Control of gene expression in eukaryotes is also carried out at the translation stage, for example, through the impact on the translation initiation factor. This explains why, even when mRNA is present in the cytoplasm, synthesis does not necessarily occur.

12. The eukaryote genome is partially redundant. For example, only 3–5 % of the 35,000 genes function in a human being at the given time. Furthermore, eukaryotic genomes contain sequences that are repeated dozens, hundreds, or even millions of times. Repeated sequences carry out various biological roles: regulation of DNA reproduction, participation in crossing-over, definition of borders between exons and introns, etc. Among repeated sequences, there are elements with changeable location, called transposons, or mobile elements.

13. Regulation of expression can be carried out at the stage of post-translational changes. For example, in the formation of the active form of the protein hormone insulin, two chains are cut from the proinsulin molecules, which are then stitched together in a different way by means of disulfide bonds.

14. Not just one, but many genes participate in the synthesis of one protein. In particular, these include one or more structural genes, delite which form mRNA; approximately 20 genes of tRNA, which are required for transfer of the 20 different amino acids (protein monomers); and several genes of rRNA for the formation of ribosomes. Many other structural genes of various enzymes for processing, translation, modifications, etc., also function in this manner.

15. In the process of ontogenesis of multicellular organisms, differential gene expression is observed. That is, different genes function at different stages of embryonic and post-embryonic development, providing gradual, accurate, step-by-step development of an organism. This ensures selective differentiation of cells and parts of the developing organism. Genes that maintain integrity of the body are expressed during life and functioning. Special groups

of genes that maintain individuality of these body parts are expressed in various differentiated tissues and cells. Thus, it can be concluded that the life cycle of an organism is a cycle of differential gene expression of a rather flexible genome.

16. *Emergent embryonic expression* (EEE). Embryonic development is based on emergent expression of the genome of a zygote. Since it is known that embryogenesis repeats phylogenesis, and a developing embryo goes through all the stages of morphogenesis, it is clear that all the required information is located in a genome. The necessary information was gathered in the genomes of the ancestors of mammals for billions of years, while expression during embryogenesis takes place in just a few days. This is a genuinely explosive process of matter transformation that compresses several billion years of transformations into several dozen hours. In this respect, rather like in a kaleidoscope, the forms and states of cell systems change immediately one after another.

17. *Differential expression of the Integrated Global Genome.* As already noted, the whole set of genomes of all living organisms on Earth can be presented as an integrated GG system. Then, it is evident that all resulting phenotypes stem from its differential expression and constitute together an integrated Global Phenome system.

Clearly then, the processes expressing genes and genomes in traits  and phenotypes are the most important mechanisms of life. In essence, these very complex catenary processes for materializing information ensure the germination and development of bodies that maintain the phenomenon of life.

## 22.4  Genomes

Recently, when we began to understand the integrity and complexity of the genetic apparatus, this term was taken to refer to the totality of sequences and segments of DNA of a specific karyotype that are structurally and functionally tied into an integral system containing a complete set of information about properties and processes of a living body, as well as mechanisms for its extraction and transformation.

The main informational elements of a genome are genes. The quantity and quality of functioning genes in multicellular bodies depend on the type of tissue a cell belongs to, the stage in its life cycle, and the stage in its individual development.

The genome also has mobile elements, for example, transposons, segments of DNA molecules that carry information about proteins participating in the transposition of this area, or episomes, which are foreign gene sequences that have been included in the DNA of a host and replicated along with it. There are also regulatory areas that bind signal molecules, segments that intensify or weaken the action of genes, and also a large number of elements of the genome whose functions have not yet been understood.

In addition, all animals also have DNA in mitochondria. They contain only a few dozen genes and other segments, but make an essential contribution to the energy of cells. Such "mitochondrial chromosomes" are a part of the integrated system of the genome. (Plants also have DNA in chloroplasts.) It is also important to add the totality of enzymes and various regulatory factors that serve DNA. It should not be forgotten that all this is located in a highly organized colloidal matrix of protoplasm. In other words, as a structural and functional unit, the genome is not just a complex of DNA molecules, but also the protein composition of chromatin, special packaging and spiralization, hundreds of enzymes and factors, and a highly organized surrounding environment. All the elements of this dynamic system are in constantly controlled interaction. Therefore, rather than saying 'genome', it would be better to use the term 'genome system', because it better reflects the essence of this highly organized structure.

Thus, the genome is part of the cell which contains a specific set of NAs and proteins combined into a single structure-functional system. This system contains special genetic information, as well as mechanisms and tools for its application. A phenome is part of the cell that surrounds the genome and integrates it into itself to form a monolithic body. The phenome of a multicellular organism is composed of an organized mass of cells and intercellular substance, so it is a highly organized colony of standard genomes within the phenotypic framework.

*Human genome.* A human karyotype contains 46 chromosomes. Every interphase G1 chromosome contains one molecule of DNA comprising a large number of genes. Genes are located in a linear order. Every gene has its place, or locus. The genome of a person contains over $3.2 \times 10^9$ nucleotides, sufficient to form a two million genes. However, research shows that a human organism has approximately 35,000 genes. Only a few percent of them are used when the organism reaches maturity. It is evident that a significant part of the genome is used for the processes of embryonic development, differentiation, and growth, and is not further expressed. Another significant portion of redundant DNA belongs to the composition of introns. An even larger portion of DNA makes up numerous families of "meaningless" repeated sequences. Such sequences may occur as often as $2-10^7$ times per cell, although they may possess as yet unknown functions.

Different genes may be organized in different ways in DNA. For example, the same genes can be repeated many times, forming tandems (for example, rRNA genes). Cluster genes are groups of different genes in a specific chromosomal area that are united by common functions. In particular, clusters of five different histones are repeated 10–20 times. Solitary genes among satellite DNA generally provide regulatory or intensifying action regarding structural genes, as exemplified by enhancers.

The majority of DNA (up to 95 %) is located in the nuclei of chromosomes. Approximately 5 % is mitochondrial DNA. Mitochondrial double-stranded DNA (mtDNA) is called the 25th human chromosome. In every somatic cell, there are hundreds of mitochondria. Mitochondrial DNA is replicated and transcribed autonomously from nuclear DNA. A minor amount of genetic material consists of small annular molecules of DNA located in the nucleus and cytoplasm.

Chromosomal DNA of a nucleus is divided into two groups depending on the nucleotide composition of segments: (a) with unique sequences of nucleotide pairs which contain the majority of genes; and (b) with repeated "meaningless" sequences. Areas with repeated sequences differ in the length of every repeat and the number of such repeats. If repeats consist of 2–8 pairs of nucleotides, they are called micro-satellites. Another group of repeats varies between 10 and 100,000 pairs of nucleotides, sometimes even more. These repeats are called mini-satellites. There are moderately repeated sequences (up to 1,000 repeats in one locus) and highly repeated ones (over 1,000 repeats). They can be in one locus or in many loci of one or different chromosomes. The same succession can be repeated in various loci for a different number of times. Such repeats are called tandems. Mini—and micro-satellite tandem nucleotide repeats are spread around a genome and constitute a combination of repeats unique for every organism as described by the number of repeats in various loci or in a number of loci. Their presence characterizes genetic polymorphism for each individual, and it is used for medical-genetic and court-medical purposes, as well as a passport system for animal breeds. However, the functions fulfilled by this portion of the genome have not yet been determined.

Polymorphism variants of the coding portion of DNA are also encountered in exon and intron sequences of molecules. These changes may be qualitative, if they are conditioned by the replacement or loss of nucleotides, or quantitative, if the number of nucleotide repeats can vary in a specific locus. The main type of genetic polymorphism is single nucleotide polymorphism. These are variants of DNA sequences in different people with changes in only one pair of nucleotides. Differences in one pair of nucleotides can occur in any DNA area: exons, introns, intergenic gaps, and repeats. They arise by mutation. In one person, 1.42 million such differences were identified. Same sex people are 99.9 % identical in their nucleotide sequences, and only 0.1 % of differences in nucleotide sequences were conditioned by phenotypic variants of individuals. The significance of SNP is well illustrated by the example of sickle-cell anaemia—a human genetic disease. The reason for this disease is modification of just a single nucleotide in a triplet, which results in synthesis of a polypeptide with just one modified amino acid. However this has a dramatic consequence: the formation of haemoglobin with abnormal spatial structure, modification of its affinity to oxygen, abnormalities in the shape of erythrocytes, etc., all of which lead to a serious pathology.

The dimensions of human genes and the number of exons and introns in them vary widely. The majority of DNA genes in a human being contain from 1,000 to 50,000 pairs of nucleotides. The number of introns in them ranges from 2 to 50.

Many genes and gene families have been identified that play an important role only in early embryogenesis. Several genes concerned with embryonic development in humans are homologous with other species of mammals. The majority of these genes regulate the production of proteins called transcription factors. They control RNA transcription, activating or suppressing the expression of genes.

The most important transcription   factors control many genes that coordinate embryonic processes, e.g., the segmentation of bodies, embryonic induction, migration and differentiation of cells, etc.

It has been established that some areas of DNA are able to shift inside a genome. Such areas are called transposons or jumping genes. They can be transferred inside a chromosome or "jump over" to others. They can place themselves in the middle of a gene, impairing its work, or join to the side, modifying its regulation. Some transposons act as enhancers, silencers, or terminators of transcription. As a result of such shifts, new genes can form, and at times even new traits  can appear in an organism. Moreover, mobile cellular elements in the form of plasmids or viruses can escape the control of a cell and transport genes or groups of genes from some cells to others, from one species  of living organisms to others.

A considerable proportion of the genome of a human being (up to 8 %) also contains a sequence of nucleotides that are offspring of viruses—*pro-viruses*. They may have appeared because of infection of embryonic cells (millions of years ago) in some primates, and since then been transferred to later generations. Many genes in human DNA came from bacteria. They are called DNA transposons. The majority of these ancient genes are "quiet." However, under certain influences on a genome, they can be activated, and this can damage the cell metabolism, in a similar way to mutations.

A complete set of genes and segments of a genome in cells and organisms form an extensive interrelated network with a multitude of feedback loops and a multitude of interactions. In this context, practically all genes can directly or indirectly regulate each other's activity or function mutually. However, in spite of significant achievements in studies of the genome, the principles of organization and operation of this complicated system have not yet been fully understood.

Thus, a genome is a constantly functioning complex system. The possibility of transfer of nuclei attests to its relative autonomy and primary role. The main components of a genome are NA and hundreds of special proteins. Taken separately, neither proteins nor NA manifest any property of the living. Only their symbiosis in a water-colloid environment can exhibit new structural and functional features. Based on the autonomy, integrity, and ability to reproduce, as well as the ability to create phenotypic surroundings, the genome can be considered as the reason for the existence of living bodies and the means of their maintenance.

# Chapter 23
# Functional Systems of Genes

## 23.1 Dynamic Genetic Nets

It has been established that the units of expression of genomes in various traits are not separate genes, but their systems, or genetic networks. An example is a system of genes that encodes enzymes of a certain metabolic process. Here, the key regulator of this system immediately activates the transcription of an extensive cassette of these genes, which provide all biochemical reactions with enzymes and regulators. As a rule, the genetic network that controls specific traits of an organism includes from several dozen to hundreds of genes. Many such genetic systems can function simultaneously in the genome.

Consider, for example, a genetic network that provides differentiation and maturation of erythrocytes launched by erythropoietin. Binding with a cellular receptor, it activates a protein kinase path for the transfer of a signal into a cellular nucleus. As a result, transcription factors are transferred into the nucleus and activate the transcription of a number of genes, including a gene that encodes a GATA-1 factor, which is the key to the erythrogenesis network. This binds in regulatory areas of all the relevant genes, including those in a promoter of the actual GATA-1 gene. As a central regulator of the genetic network, GATA-1 provides simultaneous activation of a large group of genes, necessary for erythrocyte differentiation. In this case, the significant role of positive feedback loops is displayed, guiding the controlled parameter as required. They play a key role in genetic networks for growth and differentiation of cells, morphogenesis of organs, and growth and development of organisms, where biosystems are in constant development.

The genome of any organism contains a great variety of genetic networks and systems that control molecular, biochemical, physiological, and morphological traits. In every genetic network, there are genetic mechanisms which ensure its regulation by changing the functions of one or the other groups of genes constituting the network.

G. Zhegunov, *The Dual Nature of Life*, The Frontiers Collection,
DOI: 10.1007/978-3-642-30394-4_23, © Springer-Verlag Berlin Heidelberg 2012

Regulatory areas of genetic networks in eukaryotic DNA are huge. They may be significantly bigger than coding parts of genes. These areas can comprise dozens of regulatory elements. For example, the regulatory area of a gene for rat tyrosine aminotransferase contains 40 regulatory units—binding sites for various transcription factors. The length of this area is approximately equal to 10,000 pairs of nucleotides, which is 10 times more than the coding portion of the gene. Thus, in eukaryote genomes, there is plenty of allowance for regulation. This contributes to the fact that the same gene in a multicellular organism can function in a different manner in different cells, tissues, or organs. A quite specific set of regulatory transcription factors can form in the nucleus of a certain cell in response to various impacts of the external environment, depending on the situation in the cell or the action of various inducers. Connecting with special sites within a regulatory area of a gene, they form one or the other variant of a transcription complex, providing the right expression for this gene.

As a rule, genetic networks are arranged in such a way that each has one central regulator—a transcription factor that activates a large group of genes at the same time. The systematic activation of large groups of genes has great functional significance, because they encode a similarly large group of proteins which, at the same time, should be present in cells in stoichiometric quantities to carry out molecular-genetic or metabolic processes. For example, when differentiating an erythrocyte, the GATA-1 factor activates the genome system that encodes all the enzymes ensuring the synthesis of a GEM, as well as alpha- and beta- subunits of haemoglobin, i.e., the whole set of molecular components required for formation of haemoglobin molecules.

If mutation takes place in the central regulator area, it changes the work of the whole complex of genes it controls. This type of system mutation can simultaneously cause coordinated changes in the functions of many genes and, as a result, modify many phenotypic traits. These very same mutations have significant evolutionary potential. Most likely, new morphological and physiological systems have arisen on this basis.

There are also so-called genetic network integrators. These genetic networks play an important role in integrating and coordinating the functions of local genetic networks, responsible for implementing specific target functions. Such networks receive external activating stimulus through the reception system, transform it into various specific forms, and distribute signals through their exits, connected with other genetic networks, thereby activating them. In their turn, each of these genetic networks transfers the obtained signals to other genetic networks connected with it. In other words, genetic network integrators are key elements of the genome, distributing an excitation throughout the whole organism.

In any genetic network, an executive component can be determined, which ensures implementation of a basic process, and a regulatory component, which adjusts active processes to the required level of intensity, depending on the internal condition of an organism and parameters of the external environment. As an example, consider a tricarbonic acid cycle. An executive component includes the provision of 139 metabolic reactions. However, it is interesting that the regulatory component provides a significantly greater number of regulatory reactions, viz., 1,882. In other words, there are on average 13 regulatory reactions for each metabolic reaction. This means that the regulatory component of a genetic network is much more complex than an executive component.

Executive components of genetic systems of basal metabolic processes were formed billions of years ago. In essence they are similar for representatives of various global taxons: archobacteria, eubacteria, and eukaryotes. However, regulatory components of basal metabolic processes differ significantly, even at lower taxonomic levels. This means that during three billion years of evolution of life on Earth, the targets for evolutionary transformations were above all the regulatory systems of organisms.

Regarding executive components in metabolic systems, these have remained practically unchanged during a long period of evolution. In particular, the number of genes in the majority of studied eu- and archaeobacteria is approximately 2,500–4,500 genes (2.2—4.6 million nucleotide pairs). This is clearly the minimum needed for survival and replication. Approximately the same number of genes is found in the simplest eukaryotic organisms. For example, in unicellular yeasts, the size of the genome is 12 million nucleotide pairs, which corresponds to around 6–7 thousand genes, close in value to bacterial cells. Thus, in elementary eukaryotic organisms, in spite of the presence of a nucleus, diploid ability, a complex system of biomembranes, cytoskeleton, etc., there was no quantitative or qualitative increase in genome complexity compared with prokaryotes, judging by the number of genes. It is more likely that essential phenotypic differences were caused by mutations at the level of the regulatory components of genomes.

A qualitative leap occurred in the complexity of molecular-genetic systems (as evaluated by the number of genes) with the appearance of multicellular organisms. For example, an annulate worm *Caenorhabditis elegans* is one of the simplest multicellular organisms, containing only about 1,000 cells, but it already has about 19,000 genes. As an order of magnitude, this is close to the number of genes in a human being, which is 35,000 genes. A fruit fly has approximately 12,000 genes, and angiosperm plants have 15–25 thousand genes. Fish, amphibians, reptiles, and mammals also have similar numbers of genes in their genomes to within an order of magnitude. However, the phenotypic characteristics of the above-mentioned organisms differ essentially. This attests to the fact that, in multicellular organisms, there is no direct relationship between the number of genes and the phenotypic complexity of an organism. Such a tremendous difference is more likely to be connected with the changeability of the regulatory systems of a genome within the genetic networks and regulatory systems of an organism.

Genetic functional systems are very dynamic. That is, some genes can function as part of different genetic networks. This is exemplified by the genes of different transport and ribosomal RNAs. Depending on regulatory signals, the system may include various groups of genes. This allows for the expression of a significantly greater number of features in relation to the number of available genes. In other words, both the number and the quality of features are determined not only, and not that much, by separate genes, but rather by their possible combinations and interactions in genetic networks. As an example, since a human genome has approximately 35,000 genes, their possible combinations in genetic networks is effectively an infinite set. Thus *Homo sapiens* can possess a corresponding multitude of macro- and micro-features and their combinations.

From what has been said above, it is also obvious that there are no specific genes for complex traits. For example, there are no such genes as an ageing gene, an intellect gene, a talent gene, or an obesity gene, i.e., a gene for a specific function, property, or ability. There are also no specific genes that would determine whether a given individual belongs to a certain species—there is no gene for a man, a chimpanzee, a cat, or a mouse. All these complex sets of traits are determined by functional systems of genes that are specific in each case—genetic networks that include dozens, hundreds, and thousands of genes.

Some traits are similar for various species, or even phyla or kingdoms of living organisms. For example, the process of glycolysis is inherent to practically all living organisms. It shows that the same genetic networks can function in various very different organisms, despite their completely different genomes. Moreover, the multitude of species of living organisms is determined not only, and not actually very much, by the qualitative and quantitative composition of genes, but rather by variants of their involvement in functional systems. In other words, a genome of a certain species is a finely adjusted functional system. The external manifestation of this adjustment is, for example, characteristic of a karyotype: the number of chromosomes, size and form of chromosomes, availability of sex chromosomes, etc. Internal adjustments are, first and foremost, the availability of specific regulatory segments of DNA. Thus, genes do not function separately, but rather they are expressed as whole groups, united into functional systems.

A set of genomes of individuals of the same species can also be considered as a specific genetic network. This network is a component of the overall genetic network—the Global Genome system. In its turn, the Global Phenome system is formed and maintained through the expression of the GG, and this includes the totality of individuals and their genetic networks.

## 23.2  Gene Transfer

Transgenesis is the transfer (See also 10.1) and integration of genes of certain organisms into various DNA segments of other living bodies. Such a phenomenon is possible because genomes of absolutely all organisms have the same nucleo-protein nature. Using special methods, gene engineers can persuasively demonstrate the possibility of horizontal genetic information transfer.

Transgenesis occurs constantly in Nature. The simplest example is an infection of an organism by some virus. This process can be considered as genetic aggression of viruses against the cells of a certain organism. NAs of viruses permeate cells where, using special enzymes and processes, they can integrate into foreign DNA. Then, the transgenic segment can be expressed along with genes of the host cell, which ensures reproduction of the virus. Alternatively, it can exist as an episome in a hidden form for a long period of time.

Another example of natural transgenesis is the ability of the soil bacterium *Agrobacterium tumefaciens* to induce tumors in plants. This is connected with the transfer of a plasmid into the genome of plant cells, and it is nothing but transfer of genes between the representatives of different kingdoms of living organisms. There are scientific publications on the transfer of genetic material between the cells of mammalian organisms, which is supported by the expression of marker genes in recipient cells. The ability of DNA segments to permeate even through the barriers of the digestive tract has been demonstrated in research which shows some marked areas of nucleic acids of genetically modified food found in the tissues of these animals.

Several decades ago scientists learned to create conditions for artificial gene transfer from one organism to another. For this purpose, a set of gene engineering methods was developed. It includes synthesis of specific genes, splicing of specific genes from DNA, gene identification, obtaining plasmids (carrier vectors), integrating genes into plasmids, transferring genes into a foreign genome, etc. These methods led to the possibility of genetic therapy for some hereditary genetic diseases in humans, diagnostics of molecular diseases, and the development of transgenic animals and plants. Transgenic organisms are creatures that contain foreign genes in their genomes, these being deliberately transplanted using the methods of molecular biology. Currently, quite a large number of such organisms have been created. In particular, there are transgenic mice, rams, fish, and other animals. They have foreign genes in their genomes that condition the presence of new features, useful for humans.

Billions of years ago, transgenesis was probably a very common phenomenon for primary forms of life. Like mutations, it was also a powerful factor of evolution, since it resulted in the enlargement and diversity of genomes, which in turn conditioned the emergence of new phenotypes. Natural selection fixed the better adapted organisms, and this gradually led to the appearance of new species. Even today the transfer of genetic information plays an important role in the evolution of prokaryotes that have special mechanisms for this purpose: transduction, sexduction, transformation, plasmid transfer, etc. (Fig. 23.1).

It seems likely that, at a certain stage of evolution, cells started "producing" viruses and other derivatives of their genomes. Plasmids, viroids, viruses, and then episomes, IS-elements, and transposons became a means for exchange of genetic information. These mobile genetic elements began to participate in the processes of changeability and evolution based on the general polynucleotide systems. In other words, viruses and their analogues were formed as a global molecular mechanism for transfer and exchange of genetic information between genomes of various cells (viruses, viroids, phages, and plasmids), as well as inside an individual genome (transposons and IS-elements). They joined processes of horizontal transfer (from one genome to another, from one organism to another) and vertical transfer (replication, transcription, translation, expression) of genetic information into the integrated network where bio-information circulates inside the integrated system of the Global Genome (Fig. 23.2). It can be assumed that mobile

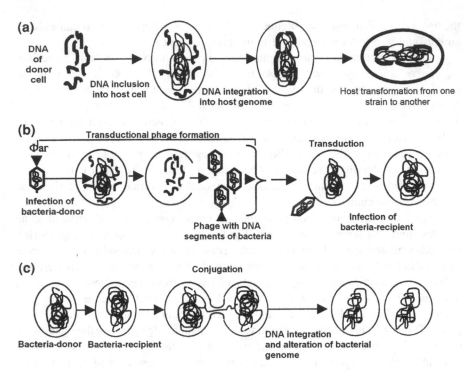

**Fig. 23.1** Transformation (**a**) transduction (**b**) and conjugation (**c**) are the ancient mechanisms of transgenesis and transmission of information

genetic elements constitute the mechanism that links discrete genomes on Earth into an integrated informational field. Then, any mutations and recombinations of a genome of any organism become the heritage of the whole GG system and can be manifested in various highly unpredictable forms of phenotypic frameworks. Moreover, any gene or NA segment can be transferred to any part (any genome) of the integrated informational field.

Thus, all the living matter is interconnected by flows of bio-information, which circulates within the GG system. The presence of an integrated information space and the global interdependence of all information carriers provide a global perception of genetic fluctuations, their transmission to any parts of the GG, and their manifestation in unpredictable forms. Such fluctuations of genetic information may be one of the most important mechanisms of evolution, and viruses and their analogues are effective tools in an infinite process of evolution.

Horizontal and vertical exchange of genes has resulted in the genetic integration of all living organisms. Due to billions of years of uncontrolled horizontal transfer of genes, prokaryotes have virtually no endogamous 'biological species'. Thus, microbial colonies may be considered as integral organisms. The whole global prokaryotic biomass may be considered as a giant polymorphic superspecies. Eukaryotic biomass, due to various mechanisms of survival and adaptation, has

**Fig. 23.2** All sorts of information transfer are possible in the Integrated Information Space of the Global Genome. Horizontal and vertical transfer of genes occurs within each kingdom of genomes. And the totality of GG information is united into an integrated system of circulation

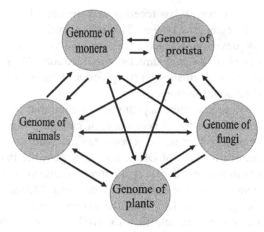

INTEGRATED GLOBAL GENOME

developed genetically and phenotypically separate groups of organisms—relatively endogamous classes and species. However, despite the phenotypic diversity, the whole prokaryotic and eukaryotic biomass of the Earth is nothing but the integral Global Phenome, which is the derivative of the Integrated Global Genome.

Moreover, genomes, and phenomes are linked into a single system, not only through the circulation of genetic information, but also through the genetically determined interdependent behavior of living bodies, which is in accordance with Richard Dawkins' concept of the *extended phenotype.*

Thus, genes can be transferred, integrated, and expressed in "foreign" genomes. Transgenesis demonstrates the tremendous potential of genetic apparatuses to undergo very fast and very significant modifications, which can cause further dramatic changes in phenotypes, their properties, and characteristics. Transfer of genes within the integrated information space of the GG, together with mutations and recombinations of DNA, may explain evolutionary leaps in the development of living nature.

## 23.3 Directed Changes in the Genome

The synthetic theory of evolution considers the genome as a passive structure that encodes and transfers incidental mutations and recombinations of DNA. It is generally considered that only natural and artificial selection play any active role, selecting and securing useful modifications of a genome into adequate adaptive traits. In other words, the genome can react to information that it receives from the external environment, but only if it follows the conditions created by artificial or natural selection. By itself, it cannot condition the directed creation of such adaptive programs.

However, it has recently become clear that the genome is a self-regulating and self-organizing system. This info-genetic system includes not only the material for the development and evolution of living nature, but also the molecular mechanisms of its own directed modifications for creation of new genetic programs, active participation in the process of its own evolution, and evolution of its own derivative phenotypes.

Recently, there appeared an opinion that the most successful genome is the one that is capable, if necessary, of changing significantly and quickly in a specific direction. That is, the genome has mechanisms that create and regulate coordinated transformations of DNA. It is quite possible that these molecular mechanisms are controlled by a cell (phenotypic framework of a genome), which is in direct contact with the external environment. Mechanisms for the adequate response of the genome to factors of the external environment could have appeared like any other function, and then evolved together with its carrier. Finally, this function would become inherent to the genomes of multicellular organisms as well. This corresponds quite well with our concept of the priority of the genome, which builds a phenotypic framework in the form of a cell or a certain body, which are in turn intermediaries in interactions between a DNA complex and the external environment.

It has become apparent that transformations of a genome are often activated at the moment when an organism needs them. One first observes cellular mechanisms of information transfer from the external environment into a genome and then molecular mechanisms of an adequate response.

Executive mechanisms that could participate in directed transformations of the genome are, for example, mobile genetic elements. It is known that transposons play a significant role in natural gene engineering. They are considered as intra-genomic molecular systems inducing DNA transformations that form the basis for various genotypic and phenotypic changes. Transposons range in size from several hundred to several thousand pairs of nucleotides. In a chromosome, they can be present in several copies. From time to time, transposons can be activated under the impact of their own enzyme and transferred to a different area of DNA within the same cell. This process is called transposition. Enzymes, called transposases, catalyze this reaction.

Mobile genetic elements combine the ability of relocation, capturing and carrying adjacent nucleotide sequences, creating retro-copies of transcribed sequences, and ectopic recombination between homologous sequences. These mobile areas of DNA participate in the creation of new regulatory elements inside their genome. The very moment when changeability mechanisms are activated is not incidental, and the spatial distribution of changes is not incidental either. However, it is still not known what triggers transposition.

Recently, a new branch of genetics and molecular biology called genomics has been under intensive development. This studies molecular organization and mechanisms of genome functioning. Comparative genomics is expected to bring clarity to our understanding of the forces and patterns that underlie the formation and development of genomes. The existence of "cold" and "hot" areas of changeability in the genome, specialized "fragile" areas of DNA, breakage and reconstruction of chromosomes, mechanisms underlying the appearance and distribution of segment duplications, irregularity of genome changeability in various types—all these will significantly complement our understanding of the evolution of the genome, as postulated by the synthetic theory of evolution. A large amount of data will be obtained by decoding the primary sequences of genomes in various organisms and carrying out comparative research, whereupon we may be able to uncover the mechanisms that underpin directed evolution of the genome.

Besides programmed reconstruction of a genome, there are probably also intercellular molecular systems that determine locus specific appearances of mutations in DNA. There may be a molecular apparatus to attract activated protein complexes to specific loci at some specific time. In other words, a special cellular molecular apparatus may connect the informational surroundings of the external environment directly with genome. This messenger would determine both the temporal and spatial coordinates of the mechanisms that purposefully reconstruct a genome. From the above discussion, we can see that both functioning and reconstruction of the genome are determined not only by the elements of nucleic acids, but also by dozens of different proteins and their complexes. In other words, the protein component of genome functioning and reconstruction is no less important than the nucleic component.

Several systems that control both programmed and non-programmed reconstructions of the genome can be distinguished. These are, for example, DNA elements, with which regulatory proteins can connect in a specific way; complementary interactions between RNA and DNA; and epigenic modifications of histones (e.g., structure modification) and DNA (e.g., methylation). Besides these, some other mechanisms that regulate DNA metabolism can participate in the process of reconstruction of a genome: enzyme molecular complexes that implement replication, transcription, reparation, modification of DNA, condensation and decondensation of chromatin, etc. These facts were determined when studying genome changeability of tumor cells, transposition, spontaneous mutagenesis of yeasts, etc. Thus, it is quite possible that various stress conditions assist in switching regulation mechanisms of the DNA metabolism to work towards directed reconstruction of a genome.

A form of epigenic heredity has been discovered recently, which is not connected with changes in the genetic code. For example, this concerns modification of histones, DNA methylation, changes of NA conformation, etc. Through inheritance of such changes, they can slowly accumulate, consolidate, and strengthen. Then, at some point, such modification can activate mechanisms of changeability under the impact of internal or external factors and play a significant role in directed reconstruction of the genome, creating areas of genomic lability and conditions for reconstruction of genetic material. In other words, the genome is a dynamic and rather plastic system, in which changes occur all the time. Many reconstructions take place naturally, but not because of some blind accident.

It is obvious that, for a cell which possesses unique molecular instruments for using and reconstructing the genome, it is worth applying them at the moment when changes are necessary for survival and evolution. From this point of view, mobile genetic elements (including viruses) are not parasitic DNA, but instruments of natural gene engineering. Known enzymes of mobile genetic elements, such as integrases, reverse transcriptases, endonucleases, transponases, etc., may modify a genome with the help of insertions, duplications, inversions, and transductions. Thus, mobile genetic elements reconstruct a genome, assisting the appearance of directed mutations through recombinations. Mobile elements of the genome can be in a non-active state throughout a number of generations. However, genomic stress caused by external impacts or appearing, for example, as a result of recombination errors, can immediately, or after several generations, induce transposition changes in the genome which are phenotypically directed against irritating factors.

Thus, modern research reveals that, along with the classic conception of the mechanisms and active forces of evolution, it is also important to take into consideration the possibility of purposeful genome modifications. Apparently, the processes of adaptation and evolution are not incidental, but rather purposeful. The genome is not a passive structure which encodes and transfers incidentally appearing mutations for active selection, but a self-organizing and self-regulating system that includes both information for development and evolution, and molecular and cellular mechanisms for its own directed modification and the creation of new genetic programs. In other words, the genome is primary in relation to its phenome. A host genome makes a path towards its existence under various environmental conditions, using its own phenotypic framework as an instrument of interaction with the external world.

# Chapter 24
# Genetic Continuity

## 24.1 Bioinfogenesis and Infobiogenesis

The evolution of phenotypes is based on bioinfogenesis—the emergence of new genetic information in their genotypes during the process of ontogenesis. In turn, use of this information—infobiogenesis, causes the emergence of new phenotypes and their subsequent development.

*Bioinfogenesis* is a process of generation of new information in genomes: new genes, alleles, and their combinations: initial genome → modified genome. In this case, the primaries are mechanisms of formation of new information, e.g.: 1. hybridization; 2. mutation, 3. recombination, 4. transgenesis, and 5. directed genome modifications.

In particular, new combinations of genes are constantly formed in the process of sexual reproduction. Several cytogenic processes condition this. For example, it is conditioned by crossing-over during meiosis at the stage of gamete formation. It is also promoted by multiple combinations, when the genomes of a father and a mother are united, as well as various options of chromosome separation in the anaphase process of meiosis. These peculiarities of genetic apparatuses are widely used by selectionists for nurturing new breeds of animals and plants. Through hybridization, dozens of breeds of cats, dogs, chickens, and many other organisms that contain various allele combinations of genes have been developed.

Thus, during reproduction, there is a constant shuffling of genes which conditions an endless number of allele combinations in new-born organisms. As a result of natural selection, the particular alleles and their combinations that assist survival and adaptation are consolidated in an organism. New combinations of genes are formed during asexual reproduction as well; in particular, somatic meiosis is already acknowledged as a widespread phenomenon. In this way, gradual evolution, based on reproduction and the succession of generations, brings about the formation and accumulation of genetic information.

Mutations constitute a very important mechanism for generating information. They bring about variants in the changeability of separate genes and their

G. Zhegunov, *The Dual Nature of Life*, The Frontiers Collection,
DOI: 10.1007/978-3-642-30394-4_24, © Springer-Verlag Berlin Heidelberg 2012

chromosomes, as well as karyotypes. This affects the structure and functions of cells and organs, and leads to the formation of new properties and traits in organisms.

Transgenesis is a process of inter- and intragenome circulation of genetic information, transfer, and integration of various genes (various segments of DNA) of one organism into different genomes. Various mobile genetic elements, such as viruses, transposons, plasmids, episomes, etc., condition the reconstruction of genomes and the emergence of new information.

Thus, bioinfogenesis constantly leads to the emergence of new information in the genomes of organisms, based on their modification and following fixation in the Global Genome as a result of natural selection.

*Infobiogenesis* is a process of realization of genetic information during the formation of living bodies (including information which appeared as a result of bioinfogenesis). Information is implemented by a classic protocol: DNA—RNA—protein, using the mechanisms of replication, transcription, translation, and gene expression. At the same time, an organism gradually forms in the process of step-by-step expression of the modified genome. It has modified size and shape, and also possesses individual properties and functions. Natural selection fixes phenotypic options adequate to the external environment, and thereby ensures the evolution of living beings.

Infobiogenesis also includes the emergence of new information in a developing biosystem, such as formation of cytological structures, functions, and mechanisms. It can be called *structural or phenotypic information*. Together with genetic information, this structural and functional information is a prerequisite for each subsequent stage of development. For example, everything starts from the union of genomes in a zygote—a new living system appears, whose structure and organization (not only the genome) conditions a number of targeted processes. Next, a blastula is formed. This is a new structural and informational level of the developing body. It possesses another phenotype, and therefore a new set of phenotypic information that conditions the next stage of development. Then, a new level of structural information—a gastrula—conditions the differentiation and emergence of germinal layers. This level of information conditions tissue formation. And so it goes on, until a mature organism is formed. Parallel use of both the genetic and phenotypic information of biosystems provides for their full and directed development. At the same time, genetic information is converted into phenotypic information. This means that information can go from one form of existence to another. It seems likely that phenotypic information is a necessary addition to genetic information. It provides such parameters of biosystems as size, location, growth direction, etc. In other words, it appends the information that is likely to be missing in a genome.

Thus, bioinfogenesis and infobiogenesis are two closely connected components of a general cyclic process of generation and transformation of information (Fig. 24.1), and that is what the patterns of development and evolution of biosystems are based on. The actual process of evolution, in its turn, is a powerful generator of negentropy in the Integrated System of Life.

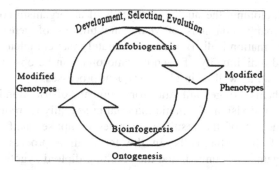

**Fig. 24.1** Creation, transformation, and circulation of bioinformation in the processes of ontogenesis, development, and evolution. Interdependent cyclic processes of infobiogenesis and bioinfogenesis are the mechanisms of the global process of initiation and transformation of biological information within the informational space of the Integrated System of Life. Modification of genotypes is a consequence of bioinfogenesis, that is, the emergence of new genetic information. And its use, or infobiogenesis, creates new variants of phenotypes, ensuring their development and evolution

The processes of circulation of information and matter described above occur exclusively through organisms. Therefore, in this context, living bodies can be considered not only as carriers of information, but also as *points and means of coupling* of special information, material, and energy flows.

## 24.2 Info-Genetic Continuity

When specifying the phenomenon of life, its properties, characteristics, and manifestations (nourishment, motion, reproduction, etc.) are usually described through observations of certain representatives of different species of living organisms. Here, life is understood as the finite existence of individual bodies. However, we understand that it is more correct to consider the phenomenon of life as a continuous process, despite the eventual deaths of its carriers. And since genomes are the main component of organisms, the phenomenon of life can be viewed as a process of continuous existence of diverse populations of genomes.

These permanent highly organized complexes travel practically without change through space and time for millions of years, using somas of various forms and sizes (various cells, organisms, and gametes), which they create around themselves. From this point of view, the various living organisms are just a means of transition with the help of which genomes move ever forward into the future.

Organisms of mammals are composed of trillions of cells. Since a zygote is the basis for the appearance of all cells of an adult organism, its nucleus is the source for the formation of trillions of nuclei, each containing the same genome as the fertilized ovule. There is a complex mechanism, based on which genetic material first doubles and then is equally distributed between daughter cells (mitosis).

Because of such cloning, the nuclei of all cells of an organism contain the same genome. Over many thousands of years and billions of generations, genetic material and information is distributed so accurately that complete preservation of the structure and individuality of each chromosome can be observed in the most complex cell cycle, as well as in all subsequent processes.

Another mechanism of genomic transformation and distribution is meiosis. This is a special type of division of genetic material in sexually reproducing creatures. The chromosome set of the resulting gametes comprises half the number of chromosomes of the initial cells. In the fertilization process, the hereditary material of the gametes is united, and the resulting diploid zygote contains a new genome. Then, and only then, this genome clones very quickly, constructing the appropriate phenotypic framework in the form of a living body. In doing so, the information in the genome is maximally realized, and the genome itself is represented in trillions of copies in the resulting multicellular body. The main objective of this formation is further production and spreading of genetic information through new gametes.

During the reproduction of unicellular bodies, it suffices for DNA molecules to replicate. They end up in different organisms right after the division of the mother cell. Unicellular organisms are rather successful. However, their genomes are not sufficiently well protected from unfavourable factors of the external environment. This could be one of the reasons for the success of multicellular organisms, where the possibility of preserving, reproducing, and distributing genomes increases significantly. However, complex multicellular organisms cannot just divide. In order for the genome not to disappear with the old organisms, it became necessary to return to the unicellular form of existence. For this purpose, Nature created an ingenious mechanism for reproduction of multicellular bodies through unicellular gametes—transitory forms of genomic existence. In the end, a worn-out organism is destroyed completely, but through fertilization and development, its genome recovers again, and a new phenome is restored. In other words, reproduction of multicellular organisms is a forced process, required for the reconstruction of "disposable soma" to provide comfortable surroundings for genomes. Thus, gametes are forced short-term forms of existence of the life substrate, as well as the means for its survival and distribution.

To illustrate the essence and mechanism of the constant development and spreading of life, we may speak of *reincarnation*. This term is used in Oriental philosophy. It refers to the phenomenon of transmigration of souls. We know that the constancy of the phenomenon of life is maintained by reproduction. Its cytological essence is the formation of gametes and their further union in a zygote. Each gamete carries a genome, characteristic for this individual. In other words, during the physiological act of reproduction and further fertilization, a transmigration of the individual genomes of parent organisms takes place. Genomes move from their bodies to new ones, joining, interacting, and creating the next carrier. If we use the philosophical term *soul* instead of the scientific term *genome*, then, along with the act of reproduction and further fertilization, the process that takes place is nothing but reincarnation, i.e., the migration of eternal souls. Continuing the analogy, we can assert that in each of us, in our every cell, there is always half the soul of our mother and half the soul of our father.

The genetic material of a human being may contain potential information about phenotypes of primates, predators, rodents, and many other evolutionarily preceding animals. It is the biogenetic law that suggests this: embryogenesis is a short repetition of phylogenesis. If in the process of development of a human being all the stages of development of the animal world are repeated—from unicellular to chordates, to mammals, we may say that this information exists in some way in its genome. So this information is in some way implemented in the process of differential expression. This may also be an example of the info-genetic continuity of the phenomenon of life.

It is important to note that, throughout the constant dynamics of life and the extreme changeability of the phenotypic expression of a genome, there are islands of strict stability, namely the practically unchangeable DNA molecules and their totality in specific genomes. Information that is read from them over and over again does not change their molecular and elemental organization, though it significantly changes the surrounding space. This is particularly obvious with the example of the development of a multicellular organism, where an extremely complicated ordered system quickly appears from the chaos of the environment. The impetus of development is the fertilization or switching on of specific genetic programs in an ovule. After that, the material chaos of the environment around the genome becomes increasingly structured as the chain process gains intensity, and after a certain time, a portion of the space is transformed into a highly organized living body, each cell of which contains a stable system of management in the form of unchangeable sets of DNA.

Thus, one can say that the phenomenon of life is the constant existence of a large number of different species of genomes, which, with the help of proteins, are able to structure the surrounding material space and determine the directions and rates of flows of matter, energy, and information. These immortal systems of genomes have travelled, are travelling, and will travel without particular changes across space and time for billions of years, using bodies of various shapes and sizes that they create around themselves.

Since the instruments and mechanisms of genomes are first and foremost keepers and carriers of genetic information, all the above can be extrapolated to its transformations. It then appears that this information circulates in space and time, using various bodies created on the basis of its own instructions.

Thus, the info-genetic continuity of Life is observed on the planet Earth. It is manifested in the form of the permanent existence of a global system of evolving genomes in the various forms of their phenotypic framework.

## Outlines of Duality or Features of Information

1.  It can be concluded from the above discussion that the dual nature of living beings as well as the phenomenon of life is associated with the presence of material and information substances. For living bodies it is the presence of the genome and its genetic information, and the phenomenon of life is characterized by the system of phenomes, which carry genomes and through which the information circulates.

2.  All kinds of movements of living bodies are controlled by genetic information in accordance with internal needs and environmental conditions. The development of the ISL is controlled by the genetically determined behaviour of living bodies and their interaction with environmental factors.

3.  Information and entropy can vary in any direction in the process of ontogenesis within individual living bodies. But in the global system of life, there is a steady accumulation of information and a reduction of entropy.

4.  Processes associated with transformation of matter and information operate inside living bodies on the cellular and molecular levels. The biological processes of the ISL occur at the ecosystem level. Material and informational interactions take place between all organisms, and these maintain the exchange processes with the environment.

5.  The molecular machines known as enzymes are the tools of the intracellular material and informational processes. Cells together with their interactions and functions are the instruments of processes within a multicellular body. The tools of transformation and circulation of matter and information within the Integrated System of Life are living bodies together with their interactions with each other and with the environment.

6.  The basis of the content and operation of the information of a cell is its genome. The basis of the content and operation of the information of a multicellular organism is its differentially expressed system of genomes. The basis of the content and operation of the information of the ISL is the interacting system of genomes of all living beings—the Global Genome.

7.  Replication, transcription, translation, and gene expression are the mechanisms that apply information in cells and multicellular organisms. Transgenesis and hybridization of living bodies are the mechanisms of circulation of information at the ISL level.

8.  Informational and material transformations in cells and multicellular organisms are implemented by the pathway DNA - RNA - protein - trait, while the information within the ISL is realized through intergenomic interactions (as indicated in Figs. 23.2 and 24.1).

9.  The functional systems for circulation of information in living bodies are realized by functional nets of genes in genomes, but also by the interactions of genomes in individual cells. At the level of the ISL, it is the interaction of species of genomes by means of hybridization and transgenesis.

10. The information within the genomes of living bodies may be modified in a directed way under the pressure of environmental factors which entail adaptive transformations of the relevant organisms. This in turn results in the adequacy of the ISL to the geophysical conditions of our planet, as a derivative of that planet.

11. Modifications of genotypes result in the emergence of new genetic information in them (bioinfogenesis). Application of this information creates new varieties of phenotypes and ensures their development and evolution (infobiogenesis). Bioinfogenesis and infobiogenesis are the main mechanisms of emergence and transformation of biological information in the Integrated System of Life.

12. Genetic continuity is inherent to the systems of genomes of individual bodies, populations, and species. It is achieved through the processes of replication, mitosis, meiosis, and hybridization. This results in the informational and material continuity of the ISL, as manifested by the permanent existence of a global system of evolving genomes in various forms of phenotypic framework.

13. The information in a genome runs the life of a body by directing the flow of matter and energy transformations into specific pathways, and by controlling the rates of these pathways and the limits of their distribution. The DNA of a genome controls the direction and rates of movement of matter by establishing certain structures and mechanisms which distinguish and guide the flows of substances and energy. The information in the GG manages the ISL and ensures its evolution by creating and maintaining countless living bodies, complementary to each other and to a variety of environmental conditions.

# Recommended Literature

1. Watson, J.D., Crick, F.H.C.: Molecular structure of nucleic acids. A structure for deoxyribose nucleic acid. Nature 171, 737–738 (1953)
2. Watson, J.D. et al.: Molecular Biology of the Gene, 3rd edn. Benjamin-Cummings, Menlo Park (1987)
3. Watson J.D.: The Double Helix. Atheneum, New York (1968)
4. Crick F.H.C.: The genetic code. Scientific American, October 1962, pp. 66–74 (1962)
5. Nurenberg, M.W.: The genetic code. Scientific American, March 1963, pp. 80–94 (1963)
6. Alberts, B., Bray, D. et al.: Molecular Biology of the Cell: Garland Science, New York (1994)
7. Berg, P., Singer, M.: Dealing with Genes. The Language of Heredity. University Science Books, Mill Valley (1992)
8. Branden, C., Tooze, J.: Introduction to Protein Structure. Garland, New York (1991)
9. Singer, M., Berg, P.: Genes and genomes. University Science Books, California (1991)
10. Vogel, F., Motulsky, A.: Human Genetics. Problems and Approaches, 3d edn. Springer, Berlin (1996)
11. Tarantul, V.Z. The human genome.—Languages of Slavonic Culture, Moscow (2003)
12. The Universal Declaration "On the Human Genome and Human Rights". UNESCO, (1997)

13. Glick, B., Pasternak, J.: Molecular Biotechnology, 2nd edn. ASM Press (1998)
14. Dawkins, R.: The Selfish Gene. Oxford University Press, (1976)
15. Dawkins, R.: The Extended Phenotype. Oxford University Press, Oxford (1982)
16. Soifer, V.N.: International Project "Human Genome". Soros Educational Journal. No. 12, (1998)
17. Hopson J.L., Wessels N.K.: Essentials in Biology. McGraw-Hill Publishing Company, (1990)
18. Timofeeff-Resovsky N.W.: Mutations of the gene in different directions. Proc. 6 Intewrn. Congr. Genet.—Ithaca 1, 307–330 (1932)
19. Kolotova T.Y. et al.: Genomic instability and epigenetic inheritance of eukaryotes. Kharkov, OKO, (2007)
20. McClintock B.: Induction of instability at selected loci in maize. Genetics.—1953.—No. 38.— C. pp. 579–599
21. McClintock, B.: The significance of responses of the genome to challenge. Science 226, 792–801
22. Joyce, G.F.: Directed molecular evolution. Sci. Am. 267(6), 90–97 (1992)
23. Ridley, M.: Agile Gene. HarperPerennial, (2004)
24. Ridley, M.: Genome. HarperPerennial, (2006)

# Part V
# Duality of Life

# Chapter 25
# Body and Intelligence Duality

Now, as never before, numerous facts indicate that life is a very complex diversified phenomenon. But from the point of view of the presence of intellect in humans, it can be divided into two qualitatively different forms.

- *Bodily life* that unites the existence of the set of various living bodies based on biological laws (that is what this book is about).
- *Intelligent life* that has a biological basis, but is connected only with humans.

This is also a specific type of duality of life forms. Between these forms, there is a gulf as deep as the one between living and non-living matter. The mind of a human being is not just a simple ability to operate information in the same way as a computer, it is rather some sort of fabulous force, whose main property is also the ability to consciously generate information in an unlimited way, to extract it from the surroundings, and to apprehend and use it for our own purposes. It is the ability of humans to transform information into energy and highly organized matter. The category of intelligence characterizes an absolutely new property of living bodies, associated only with *Homo sapiens*. This property conditions the significant independence of humans from the environment and allows a global impact on Nature.

The essence of physical life can be encapsulated by the following dynamic formula, which is based on genetic information and its material transformations: DNA $\rightarrow$ proteins $\rightarrow$ cells $\rightarrow$ organisms. Intelligent life has the same biological basis, but has in addition a lot of its own specific characteristics: self-awareness, abstract thought, culture, spirituality, science, analysis and usage of natural resources, and much more. The human mind is the next leap in the evolution of Nature, and it is as important as the emergence of life itself in the process of chemical evolution. Intellect, as a phenomenon similar by its significance to life itself, arose at a certain stage in the development of matter, has evolved along with its carriers, and continues its unabated progress. This is evidenced by the great discoveries and achievements of carriers of intellect, literally in the last few decades (in comparison with the tens of thousands of years of previous evolution).

G. Zhegunov, *The Dual Nature of Life*, The Frontiers Collection,
DOI: 10.1007/978-3-642-30394-4_25, © Springer-Verlag Berlin Heidelberg 2012

For example, advances in the exploration of space and the ocean depths, computers and computer networks, grand engineering structures, countless examples of ingenious technology and nanotechnology, mobile communications, molecular and genetic engineering, cloning of mammals, cell and gene therapy, and much more, are the fruits of intelligent human activity. The special place of *Homo sapiens* is also connected with the fact that only humans are able to inflict a global impact on Nature, build an artificial environment, cognize Nature and humans themselves, and alter Nature and exploit it for their own purposes.

Man is a conscious being. Everyone knows exactly who he is. He knows his name, nationality, kinship, the space and time in which he exists. He has a colossal social memory and constantly learns and teaches others. He knows his own organism, controls his instincts, and understands and evaluates his actions, which are not only biological, but also of a social nature. Humans have spirituality, i.e., their lives are founded upon conscience, ethics, and morality. Humans are the only animals that have escaped from hard genetic control of all aspects of their behavior. *Homo sapiens* have acquired the capacity for self-learning and self-programming. It is the only animal that possesses the non-biological heredity which conditions social evolution and cultural development. Only humans are capable of generating such concepts as the meaning of life, the purpose of life, destiny, conscience, and love. Only humans participate in areas of such "unnatural" activities as science, art, philosophy, religion, and sports. Only man is able to realize the happiness of his existence.

From the standpoint of this book on the primacy of the genome, the mind makes the human genome much more competitive in comparison to the genomes of other organisms, and this ensures its dominance. That is, the process of intelligent activity may be regarded as an effective way to struggle for existence, which is typical only of man. This is what ensures his superiority.

Elements of intelligence are also inherent to many animals, capable of evaluation, thinking, and learning. But there nevertheless exists a clear divide between the minds of humans and animals. The essence and nature of the human mind is not yet fully understood. Although it is obvious that it is based on neural networks of the cerebral cortex, conditioned and unconditioned reflexes, electromagnetic waves and fields, electric currents, molecular and quantum processes, as well as *something* else. The question of the human mind phenomenon still awaits an answer. Its special properties and characteristics should not therefore be used in determining the relatively well-studied physical life.

# Chapter 26
# Genomes and Their Bodies

## 26.1 Community of Genotypes and Phenotypes

Living organisms have a set of characteristics which we call a phenotype. The whole body of an organism, or phenome, is formed on the basis of phenotype. Organisms also possess a set of genes that determine traits—a genotype. It is a complex of genetic programs of reproduction, development, and functioning of living bodies. These programs are recorded, stored, and reproduced in genomes. That is, living organisms are fundamentally different from non-living bodies, machines, complex constructions, and other objects, not only by possessing special structure and functions—phenotypes, but also by having special programs—genotypes—for reproduction of such complicated organisms. Moreover, these programs can be duplicated many times in the cycles of DNA replication, passed on from generation to generation through reproduction, and then produce corresponding phenotypes and phenomes by means of molecular mechanisms. The presence of genomes is an absolutely necessary condition for the existence and reproduction of all kingdoms of organisms—from viruses to mammals. Thus, the genotype can be regarded as a special structure containing an encoded program, mechanisms for applying it, and everything needed for the reproduction and development of new individuals, i.e., for the formation and maintenance of a certain phenotype. All the traits of a phenotype and their combinations determine the features of the monolithic body of a phenome.

The phenome of any organism is based mainly on the proteins encoded in the genotype. That is, the basis of phenotypes and phenomes is proteins, and the basis of genotypes and genomes is NA. There is a close connection between them through the proteins and molecular mechanisms for realizing genetic information and transforming it into a character. These mechanisms include replication, transcription, translation, expression, and several others. The presence of mechanisms for implementing programs is also one of the main distinctive features of living organisms. Thus, a living organism is not only a genome and a phenome, but also a set of molecular and cellular processes and mechanisms for their interconnection.

G. Zhegunov, *The Dual Nature of Life*, The Frontiers Collection,
DOI: 10.1007/978-3-642-30394-4_26, © Springer-Verlag Berlin Heidelberg 2012

The genotype and phenotype are interrelated by processes of reproduction and development. In each cycle of reproduction of living bodies, the information of the genotype and the matter of the environment are realized in a phenome, providing the conditions for existence of the genome. Cycles of reproduction of the genotype and phenotype are repeated over and over again, thereby ensuring their continuous interdependent existence for thousands of years. This reflects one of the basic principles of the dualistic organization of living bodies—cooperative interdependent coexistence of genome and phenome through continuous copying of information.

The same duality is inherent in the phenomenon of life. The totality of living bodies of all species on Earth makes up a vast biological system, which we refer to as the global phenome (GP). The GP continuously exchanges matter, energy, and information with the environment in order to support its organization. This discrete system is constantly being disintegrated, since each individual representative has a limited lifetime. However, the GP retains its structure, and even develops due to the property of each individual to reproduce. Reproduction, along with subsequent changes in later generations, continually works against the increase of entropy within the GP system.

The Global Phenome is the product of expression of the Global Genome—the totality of countless genomes of individuals of all species that inhabit our planet. Thus, it is also a kind of open system that interacts with the environment through its own phenotypic framework. It exists on the basis of certain laws, which are not yet fully understood. This discrete system is unified by the universality of NA, as well as the principles of recording, storing, and realizing genetic information. It is integrated into a single whole through constant vertical and horizontal circulation of genetic information. The material carriers of information throughout the GG system are special carriers of the genome in the processes of reproduction, as well as a variety of mobile genetic elements. That is, the essence of the existence of the GG system is an incessant ensemble of molecular-informational processes, continuous transfer of genetic information within and between discrete genomes (circulation of information), and continuous transition of discrete genomes from one body to another. The GG and GP are fundamentally different systems, but they are nevertheless inseparable, and it is this integral inseparable couple that forms the Integrated System of Life (ISL).

## 26.2  Phenotypic and Genotypic Life

The life of organisms is a discrete, intermittent process. One generation of organisms is continually replaced by others. The replacement of generations of genome carriers is one of the most important phenomena in the existence of living organisms. During their temporary existence, individuals pass the genetic program of their kind on to future generations in the process of sexual reproduction through the DNA of gametes, or through the DNA of the body parts during vegetative reproduction. New generations, in turn, grow and mature according to their DNA programs, then produce gametes and propagate, whereupon a new round of life begins. And so it goes on incessantly, as long as there are suitable conditions for survival and reproduction of the organisms. This continuous existence of generations of phenomes is ensured by the presence of their genomes—genetic programs of reproduction, development, and survival. Underlying the "intermittent continuity" of living bodies are the systematic processes for copying NA information and proteins during reproduction and development.

The continuity of genetic material and information in the form of DNA, despite the change of generations, is one of the main characteristics of the living. This ensures the permanence of the phenomenon of life, despite the mortality of its carriers, because they manage to either reproduce or create a transient form in which life can continue through the genetic material of gametes, spores, cysts, or other compact and very stable structures that contain the basis for life—genomic DNA.

Thus, one can imagine the flow of life as occurring in two dimensions. First there is a continuous *genotypic life* that is hidden from our eyes—the set of processes constituting the dynamic existence of genomes. Then there is a visible, intermittent *phenotypic life*—the set of processes constituting the existence of living bodies. One may say that the appearance of an organism is a kind of manifestation of the creative informational potential of the genome. Figuratively speaking, one can imagine living individuals as the carposomes that appear on the body of a permanent mycelium, an "analog" of the extensive GG network.

Any living organism can be viewed from this perspective. For example, a mouse has certain well-demonstrated phenotypic characteristics, or characteristics of phenotypic life. For example, it has a certain size, shape, and weight, and it has four limbs and a tail, a certain way of moving around, a particular form of nutrition and digestion, peculiarities of metabolism and functions, specific habitat and behavior, and much more. The process of real existence and functioning of a living body of a mouse, with all its attributes and characteristics, constitutes the essence of its phenotypic life.

It is also obvious that it has genotypic characteristics. In particular it is the presence of certain NAs, genes, and proteins, and their organization in the form of a specific genome. It is the genetic information which is recorded, stored, and reproduced by the genome. These are special molecular tools for operating NAs and genetic information. It is also the processes of replication, transcription, translation, and many other molecular mechanisms that remain invisible to us. *The existence of all elements and programs of the mouse genome, along with all processes that go toward realizing its creative informational potential, constitute the essence of genotypic life.*

Genomes → proteins → phenomes: this scheme demonstrates the dualism of life and unites the two forms of its existence. Genotypic life consists of genes and proteins, along with all the molecular mechanisms, and phenotypic life consists of living bodies that are based on proteins. Their unifying molecular mechanisms are replication, transcription, and translation. These mechanisms provide the start for a chain of enzymatic, metabolic, and physiological processes that ensure all the properties and potency of living bodies.

The central role of proteins as a link between genome and phenome is obvious. Proteins are common components of both levels of life. On the one hand, the proteins provide the molecular-genetic mechanisms and processes of the genome, and on the other hand, it is also the proteins from which organelles, cells, and organisms are constructed, and which ensure their properties and functions. The discrete units of phenotypic life are living bodies, and the units of genotypic life are genomes.

Genotypic life is the foundation. It can exist even in latent forms, without phenotypic expression. For example, some bacteria have been preserved for millennia in the ice of Antarctica, or even millions of years in the salt deposits of ancient seas. The phenotypic part of their life can only begin after getting into environmental conditions conducive to the processes of metabolism and reproduction.

The phenotypic life of a multicellular body begins with the union of genomes of gametes during fertilization. Gametes, in this case, are nothing but the intermediate link between genotypic and phenotypic life. The content of the nuclei of gametes represents genotypic life, and its cellular environment is a temporary transit facility for storage and integration of genomes. Integration of genomes during fertilization is in fact a grand event. It is a qualitative leap involving materialization of information and initiation of a new level of existence—phenotypic life.

Ageing is the wearing out of living bodies as a result of work and/or under the influence of environmental factors. This process leads to changes in the structure, and as a consequence, the functions of organisms. Ageing mechanisms of living bodies as representatives of phenotypic life have been fairly well studied, while genotypic life is apparently not susceptible to this process. It is only the body that wears out and dies, but its genome and genetic information continue to exist in the case of successful reproduction and emergence of offspring, although in a somewhat modified form. A particular genome may cease to exist only if a whole species dies out.

Despite the conditionality of our approach, many fundamental properties of Nature can be viewed more broadly from the perspective of the dualism of life. For example, the reproduction of living bodies can be seen not only as a process of reproduction of a given species, but also as one of the mechanisms of continuous circulation of genomes and genetic information in the Integrated System of Life. Mitosis or meiosis can be seen not only as processes of cell division, but also as mechanisms for the "birth" of new discrete genomes and the spread of information.

Thus, the phenomenon of life has two interrelated components: phenotypic life—the process of existence of a set of organisms, and genotypic life—the process of existence of a network of genomes. These systems are united by molecular, cytogenetic, and informational processes into the Integrated System of Life.

## 26.3 Genotypic and Phenotypic Evolution

Evolution is a process of continuous gradual development, during which organisms become more complex and better adapted. This process occurs at several levels.

1. *Evolution of living beings as a tool for evolution of the phenomena of life.* Without doubt the phenomenon of life evolves progressively in accordance with the evolution of our planet. The evolutionary purpose of this phenomenon is adaptation (or, rather, maintenance of compliance) to changing geophysical conditions. The phenomenon of life is expressed through the existence of living bodies. Evolution of the phenomenon occurs through adaptation of concrete organisms. The evolutionary purpose of adaptation of organisms and species is survival. Thus, progressive evolution of the phenomenon of life, as part of the nature of the Earth, occurs through the adaptation and survival of living beings. We can say that individual evolution is a mechanism and instrument for the global evolution of the phenomenon of life.
2. *Phenotypic and genotypic evolution.* Phenotypic life is represented by the totality of all living organisms in their interaction and interdependence. Paleontology, comparative anatomy, biochemistry, physiology, and other sciences illustrate well the successive stages in the evolution of a Global Phenome. The purpose of such "phenotypic evolution" is the need for collaborative processes of adaptation to environmental conditions (evolution) and to each other (coevolution). Phenotypic evolution is conditioned by environmental factors and genotypic evolution, which occurs at the level of genetic material.

   Genotypic life is represented by a set of genomes in their interaction and interdependence. Data from genetics and molecular biology provides evidence of continuous processes that occur in the Global Genome. These are constant flows of genetic information, both within genomes and between them, both within a single organism and between organisms, and both among organisms of one species and between organisms of different species, types, and kingdoms. The purpose of this movement is the reproduction and adaptation of living bodies to the external environment to ensure homeostasis of individual genomes and homeostasis of the entire GG, which in the end ensures genotypic evolution, associated with directed modification of the genomes themselves.

3. *Evolution of phenomes, genomes, and genetic information.* Evolution of living beings is the *consequence*, while the *reason* is the continuous progressive change of our world. Since Nature is constantly changing, the existence of anything or anyone for an indefinitely long time is impossible. This determines the limited duration of existence of living beings and their species, which have *finished forms* at a certain stage of development of Nature. However, the unlimited variety of phenomes, the differing durations of their existence, changeability, and ability to reproduce condition the superior plasticity of the Global Phenome. Because of this, the GP, as a single system, is capable of

withstanding any factors, interacting with them, and developing in any direction. This determines the stability and continuity of life.

The high plasticity of the GP is also conditioned by internal causes. First of all, it is the tendency to preserve the integrity and activity of the GG as a single network of life. Around each genome a phenotypic framework is created, which is sufficient for active interaction with the environment. The specific conditions of the surrounding space lead to the specific features of the form and content of living bodies, which are intermediaries between the environment and the genome. Genomes evolve continuously in a direction that would increase the probability of survival and preservation of the bodies in which they live, and hence their own survival.

Since genomes are the tools for storage and handling of information, their survival and evolution constitutes a concurrent survival and evolution of information. Such biological information is very specific. It determines the existence of the phenomenon of life through the existence of genomes in the phenotypic framework. Thus, the goal of evolution of living beings can be regarded as the survival of a network of DNA genomes, and hence the information which these genomes carry and express.

## 26.4  Species of Genomes and Species of Phenomes

It is possible to say that organisms are secondary temporarily existing formations, the main task of which is to create the conditions for reproduction, existence, and survival of the primary substrate of life—DNA, which is organized and exists in the form of the genome. That is, the living body is a "disposable soma", once used by a genome for its own purposes. After a certain time the body wears out, grows old, and dies. But "true life" continues to exist in future generations of organisms in the form of the genome system.

We are quite confident when talking about the existence, for example, of some *species of insect* on Earth for hundreds of millions of years, as has also been proven by paleontology, although it would be more correct to say that during this time there existed a concrete *species of genome*. Its carriers have replaced each other, but the genome itself has remained virtually unchanged.

Is it correct to say that the process of existence of genomes is life? The answer is probably affirmative, since many scientists believe that viruses are alive, even though they are nothing more than separate molecules of RNA or DNA plus several varieties of proteins, which is very similar to what we refer to as the genome. Thus, individual genomes can be considered as active *genotypic inhabitants* of the Global Genome network. Or, better to say, the inhabitants of the genotypic component of life. Genomes exist only on the cytogenetic level, using the cellular mechanisms for their own reproduction and propagation. They live in the cells of a variety of living organisms and quite easily travel from one cell to another, from one organism to another, or may even exist for a long time outside the body without any manifestations.

The phenomenon of latent life and anabiosis provides evidence for the possible existence of genomes, even under conditions unfavorable to its "phenotypic framework". They could continue to exist without a phenotypic framework, but their reproduction would be impossible, for example, in an anhydrous environment. Therefore, in order to create conditions for self-reproduction, genomes build around themselves various bodies (or find such bodies), which provide standard conditions for the molecular processes needed for reproduction and survival. Thus, the phenomenon of life on Earth may be figuratively imagined as a stream of different systems of genomes through numerous species of organisms.

Proteins are "slave-robots" that fulfill all desires of the DNA. They take care of its packing, maintain structure, and provide functions and liaison with the external environment. And most importantly they ensure its ecologically clean, stable environment in the form of the contents of the nucleus and cell. Such an environment consists only of the molecules and their complexes which are required for survival and reproduction of genetic information. In particular, the nucleus has a compact, well organized, and concentrated set of chromosomes together with enzymes and structures that are there only to service it. The intranuclear environment of the genome is virtually identical and stable in all cells of a given body, despite their phenotypic diversity.

A number of processes important for the genetic apparatus occur in the cytoplasm of cells. In particular, chromatin proteins, various enzymes of transcription, replication, and reparation, are synthesized there. The synthesis of purine and pyrimidine bases also occurs there; ATP is produced in mitochondria, and so on. There are numerous other kinds of biochemical processes of anabolism and catabolism, whose purpose is at the end of the day to maintain an ordered space around the native structure of the nucleus and genome. All this is managed by the genome, which in that way controls its own environment and homeostasis. This is achieved mediately by regulation of protein synthesis, which results in the maintenance of homeostasis and controlled surroundings.

A multicellular organism can thus be regarded as a clone of identical sets of genomes that exist under standard stable conditions of the internal contents of the cell nuclei. Despite the different morphology and function of various cells, their nuclei (incubators for genomes) have virtually the same size, shape, and content. That is, the organization of genomes and their close surroundings are the same for all sets of DNA. The number of identical genomes corresponds to the number of nuclei of cells of the particular body, and can reach up to a trillion copies. (Single-celled creatures contain only one genome). Thus, we note once again that living bodies are products of the primary biological system—the specifically organized genome.

The genome is a complete set of sequences and segments of DNA in the karyotype of a certain species of organisms, which contains a complex of information concerning reproduction, development, and survival, as well as the mechanisms for its implementation, and is structurally and functionally integrated into a unified system. This highly organized system also includes a variety of chromatin proteins, various enzymes, and a colloidal matrix karyoplasm. Genomes of different complex species of organisms were formed during long-term evolution as a result of transgenesis, consolidation of several DNAs of different single-celled organisms, DNAs of various primitive multicellular organisms, and gene, chromosome, or genome mutations. Sexual processes, recombinations, and subsequent natural selection led to choosing and fixing certain forms of cells and organisms with a particular genotype and phenotype. These "phenotypic forms of framework" of genomes had to meet several requirements: (1) creation and maintenance of ideal conditions for the genome's habitation environment; (2) maximal adaptation of its own homeostasis to various external conditions; (3) provision for interaction and matter exchange with the external environment. Thus, we can say that living body forms are primarily the intermediaries in the interaction of genomes with extremely varied environmental conditions.

The DNA molecules of a genome system are unique, because only these molecules are capable of reproduction by replication. This ensures a continuous cycle of self-replication of DNA molecules and their ubiquitous unrestrained proliferation. The mechanisms of reproduction of these macromolecules also create the necessary conditions for continuous self-monitoring and self-renewal, since it is known that, during its lifetime, the DNA is continually being damaged by the impact of various chemical and physical factors, as well as wear due to permanently repeated acts of transcription. Reproduction of DNA also underpins the construction of multi-cellular bodies and their (and hence its own) dissemination.

The "phenotypic framework" of a single genome may take different forms (polymorphism) as a result of differential expression. In this way the variety of differentiated cells, tissues, and organs, as well as different forms of the body of one genome are formed. For example, very bizarre forms of existence are produced by genomes of certain species of flat parasitic worm. In particular, the life cycle of a liver fluke *Fasciola hepatica* is associated with the existence of several distinctly different living beings, living in totally different environments:

(1) The pubescent form is a flat worm that inhabits the liver of cattle and small ruminants. (2) The egg is released into a pond. (3) The larval form, miracidium, independently inhabits the pond. It actively penetrates the body of an intermediate host—a small pond snail. (4) Miracidium lives in the body of the pond snail and gradually transforms into the next intermediate stage—a sporocyst. (5) Germinal cells of the sporocyst parthenogenetically develop a new larval generation of living bodies, called redia. (6) The redia give rise to another generation of larval bodies called cercaria. They leave the mollusk and move around actively in the water. (7) The cercaria attach to the stems of plants around the pond, cover themselves with a capsule and form adolescaria. (8) If they are swallowed by animals eating the grass, they get into the herbivore's liver and develop into the adult form, known as marita. And all these different bodies are derivatives of one genome which attempts to adapt, survive, and reproduce, and for this purpose

**Fig. 26.1** Differential expression and various phenotypic products of one genome. The genome of a common liver fluke and its phenotypic derivatives. If we scoop up a bucket of pond water, it may contain living bodies which seem to be of different species. However, they are variants of the phenotypic framework of just one genome

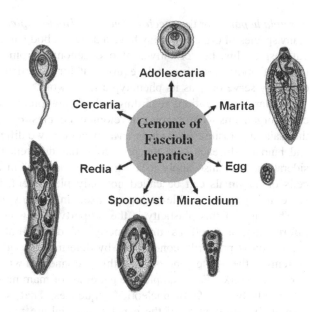

creates such bizarre phenotypes which have nothing in common except for their DNA (Fig. 26.1).

This shows that the same genome may be the source of several completely different phenomes. Thus, it is obvious that very diverse information can be extracted from the same set of DNA molecules of a genome, and a wide variety of living bodies may be constructed on its basis. This plasticity of the genome is typical for many other creatures, for example, for a large group of insects that develop with metamorphosis. We can call this property *the polyphenotypicity of a genome*, which is the cause of polymorphism.

*Monophenotypic* organisms, such as mammals, have a body which barely changes during its development and existence. In particular, the human embryo is a small body, phenotypically very similar to the adult organism, whose development mainly involves growth. *Polyphenotypic* organisms, in the process of their life cycle, have several fundamentally different bodies, as described above, manifesting completely different phenotypes. These bodies differ significantly in size, anatomical structure, functions, habitat, reproduction, etc. It turns out that the same species can exist in different bodies! So the living body is a concrete manifestation of the phenotypic framework of the genome in a particular environment and at a certain stage of development of the individual. Consequently, we can talk about the existence of a certain *species of genome* in the form of various bodies. For example, the species of genome *Fasciola hepatica* may exist in the form of a variety of living bodies: the egg, miracidium sporocyst, redia, cercaria, adolescaria, and marita. Apparently, in the taxonomy of organisms, it is more correct to speak not about the existence of the species *Fasciola hepatica* or the species *Homo sapiens*, but rather about the existence of the *species of genome*

*Fasciola hepatica* or the *species of genome Homo sapiens*. This view suggests that many species of organisms may have a different body and dwell in different media, but despite this, be derivatives of one genome. From this standpoint, all these forms of bodies are absolutely equivalent for a particular genome, since in any case they serve only as its phenotypic framework.

Similar considerations regarding polyphenotypicity also concern some single-celled organisms with complex development cycles. For example, the life cycle of the malaria parasite *Plasmodium vivax* occurs in two different hosts—the mosquito and humans. In each case the protozoan has different phenotypes. Similar considerations may also apply to a variety of mammalian cells. For example, stem cells of mammals can be called not only pluripotent, but also polyphenotypic, since their genome is able to express itself in phenotypically diverse cells.

The basis of this plasticity is the property of a genome to contain redundant information, as well as the property of differential expression. Differential expression is primarily conditioned by deterministic operation of certain genetic systems of the whole genome. And this is connected with the special regulation of genetic networks. For example, the genomes of mammals are quite large, but they differ only by 1–5 % of nucleotide sequences. That is, despite the fact that the quantitative differences of the genetic material are not very significant, the phenomes of, say, a mouse, an elephant, and a man differ enormously. Dramatic differences in the phenotypes are likely to be explained by the informational plasticity and polyphenotypicity of genomes. We may assume that many different phenomes can be reproduced on the basis of the informational redundancy of any genome.

Let us consider the following question: what is the main period of existence of a living body in a cycle of individual development? For example, what is the most important period for a chafer insect? Is it the egg, the larval stage, which lasts 3 years, or the adult stage, which lasts one month? And the liver fluke has 5 stages of development in a variety of living bodies. Which stage is the most important? A virus can exist as an autonomous body or as an episome—which is the most important? Slime molds can exist as a multinuclear plasmodium, or as many single-celled creatures. Which is the main stage? It seems that it is impossible to answer. But all of these cases lead to a common denominator, if we consider the genome as an integral and principal part of all these stages and states. That is, all the visible bodies are only the forms which hide the genome. And in all cases, the genome is the protagonist.

To the considerations presented above, we may add that the body is not eternal, while the phenomenon of life in the form of the existence of genomes is continuous and consistent. That is, the genome is practically immortal, although it can if necessary kill some of its own copies (apoptosis). In nerve cells, the human genome can live 120 years without a division, in stem cells, which constantly divide for just as long a time, and in epithelial cells where it can live for just a few days. Hence, the genome of somatic cells can in principle live indefinitely, but it is not necessary in a situation where there is a special germ line cell, incubated specifically for reproduction and distribution. The body is just a home space for a genome and its breeding, and a permanent monitoring mechanism for variability, heredity, and evolution. Through bodies, genomes enter into contact with the external environment, receiving matter, energy, and information, thus ensuring their homeostasis and reproduction. We therefore believe that the living body is only a carrier, not the main object of life; not content, but form; not the cause but only a consequence.

## 26.5 Immortality and Indispensable Death

The conscious attitude of a human being with regard to life leads to the subjective perception that this life is the chief value of Nature. This greatly complicates our understanding of its naturalness and simplicity. If we look at all other living organisms, it is clear that neither worms, nor insects, nor birds (not to mention plants and fungi) attach any special meaning either to life or to death. These biological robots live according to a strict genetic program, avoid danger, survive, and reproduce. But all living beings are submissive and indifferent to death. For example, many social insects (bees, ants), easily sacrifice themselves for the good of the family. They blindly follow the orders of their genome, as if realizing that the loss of a million DNA clones among many trillions has no meaning for the safe existence of the whole species. The interests of individuals are sacrificed for the sake of the prosperity of the community or population of a particular genome. Thus, the life of an individual organism has no specific value for a stable population or species.

The life of a mammal, as the existence of an autonomous organism, starts at the moment of fertilization—the fusion of parental haploid genomes. This forms a diploid zygote, which is a single-celled stage of the future complex organism. Then, in the process of ontogenesis, an organism undergoes several stages of development. The process of birth is important, but only a secondary phenomenon, after the germination of a new organism. (People should celebrate, besides the birthday, the more significant date—*the day of germination*, which is nine months earlier). Development stops when puberty is reached.

The essence of the above process is the reproduction of a particular genome, and its ordered distribution and packaging in a certain volume, which has the shape of a living body. Then other genetic programs begin to act. First and foremost, these maintain the integrity of the organism (colonies of genomes, embedded in a cellular matrix) and produce gametes (transient form of existence of the genome). The body produces gametes (autonomous genomes), multiplies, grows old, and after a certain period of time dies. The life of an individual organism is complete. But if offspring are left in the process of reproduction, the existence of its genome is not terminated. That is, only the carrier dies, while the genome, the main substrate of life, continues to exist in future generations. In this way, it is repeated an infinite number of times.

Thus, *at the moment of fertilization, life does not emerge as a phenomenon.* What occurs is just a transition of the substrates of life, genomes, from some individuals to others, whereupon new living bodies appear. At the same time, the organization and order of the cells cannot be said to appear again, because they were already present in the ovule and then remained in the zygote. So it is not only the genome that is being copied in the continuous process of changing generations, but also its environment in the form of the highly organized colloidal matrix of the cytoplasm. And then, by repeated cloning of the genome and cellular order, a new complex multi-cellular organism appears, carrying a gene like "the baton of life", ready to pass it on to the next generation. Thus, individual organisms are just links in the endless chain of life of a given species, while a standard genome with its highly organized phenotypic environment is a general substrate of life, which exists permanently and unites millions of generations of individuals.

We may say that the life of an individual organism is one of the phases in the cyclical development and concrete phenotypic expression of the potential of a given genome, inherent to a certain species. The beginning of the life of an individual is connected with the beginning of operation (implementing potency) of a particular genome, and the cessation of life with the end of this process in a particular body. Thus the life of the individual is a naturally determined stage in the life cycle of a particular genome, the process of implementing its information potential, surrounding itself with a phenotype, and maintaining it for the purpose of its own reproduction and distribution. And this genome and its body are in turn discrete parts of the Integrated System of Life. So we may say that the phenomenon of life is conditioned, not by existence of living bodies which have genomes, but rather by the existence of genomes, which have a bodily framework.

Thus, after the death of an individual, life as a phenomenon does not stop, if this individual has managed to produce offspring. The substrate of life, in the form of a collection of DNA, is being passed to descendants during reproduction and continues to live on in their bodies. During the next life cycle of the somewhat modified genome of the new individual, it will again be passed on to the new generation, thereby maintaining the genetic continuity of life. In short, only a particular body (discrete unit of life) actually dies, while the existence of a discrete genetic unit of life continues in other bodies. Therefore, the genomes of certain species (species of genomes) actually live many thousands and millions of years (and generations), while each body of a multicellular organism can be considered as a polygene module of a large colony (population) of the species of genome.

Thus, from the point of view of the dualism of life, the phenomenon of life is *not* exposed to ageing and never dies. It travels from one mortal body to another in the form of immortal genomes, continuing to exist for billions of years, through sequentially changing living systems.

Phenotypic life in the form of living bodies is not eternal. It is subject to ageing and mandatory death. This is only a temporary shelter for the genome, where it is cloned, and with which it multiplies and spreads. In these "disposable somas", there is obligatory systematic inspection of a genome and, if necessary, its restoration. In these bodies, there is periodic recombination of the hereditary material. The genome acquires new alleles and their combinations, without being fundamentally changed. Thus, bodies are just a very convenient transitory form of life, and the limited duration of their existence (the duration of a life cycle, from one day to hundreds of years) is not essential to the eternal process. It is sad to realize that one is only a representative of a short phenotypic life, a disposable soma of an endlessly ruthless genome. On the other hand, we can be proud of ourselves (our body), if we have fulfilled our natural duty: multiplied, and transferred our genome to descendants. This fact fully justifies our temporary existence and fills an individual life with a great sense of purpose.

Body-mediators have finished forms and properties, which prevent them from adequately following and responding to the changing conditions of the environment. Therefore, they are doomed. The owners of these bodies must modify them in accordance with the conditions of their potential habitat. To do so, they just need to change the information flow a little bit. For this purpose they have special mechanisms for their own directed reorganization, such as crossing-over, transduction, transformation, inversions, duplications, and others. Informational microchanges become significantly amplified in the process of expression and, as a result, the body is constructed with a variety of features and options. Natural selection does its own work by destroying phenotypes which are not relevant. As a result, the winner of this struggle is the permanent genome, well settled in a renewed body. Clearly, the expedience of evolutionary transformations is not the survival of particular organisms and species, but the survival of its inhabitants—genomes. Thus, the death of individuals is just one of the mechanisms of the eternal life phenomenon.

# Chapter 27
# Bodies and Associated Phenomena

## 27.1 Primary Qualities of Living Bodies

Life is necessarily associated with a certain *material body*: as a minimum, with the body of DNA or RNA; as an optimum, with the body of a cell; as a maximum, with multicellular bodies that contain from hundreds to billions of cells. Living bodies are fundamentally different from inanimate objects, not only by having a special form and content—*a phenome*, but also by having a program, *or genome*, for production of complex organisms, similar to their own kind. Proteins are the basis of the phenome, and NAs are the basis of the genome. These bodies have a discrete and highly organized structural foundation, and they possess the properties of living matter. Vital properties are not possessed by individual molecules of DNA, RNA, or proteins, but only by their system—the genome, which must necessarily be integrated into a highly organized colloidal matrix of karyoplasms or cytoplasm. It is this matrix that provides and controls precise, directed flows of substances, energy, and information through special molecular channels, and also manages their interaction. In turn, the genome, as a biological microprocessor, manages the structure and work of the molecular matrix and the entire biosystem indirectly through proteins. The cellular matrix is a system unit, and the genome is the memory and software of the entire biosystem. Separately they are inanimate but united in a single system, they acquire a new quality, which is called life.

The life of bodies necessarily involves complex, driven, interconnected *information-genetic* processes. Genes and their systems are the main units of information and control. The basic genetic matrix processes are replication, transcription, and translation. The emergence of these complex processes is even more mysterious than the emergence of NAs. Management of biological systems and their development is based on differential gene expression. Genomes ensure their own continuity, stable environment, and protection from external factors by realizing the information of the genetic system through matrix processes. They create a living body around themselves. Living organisms are "clots of information"—materialized genetic information. They are built on the basis of information, exist in a dense informational environment, and live and survive through the ability to generate, perceive, analyze, and use information. The main informational difference between living and non-living bodies is that they possess a *thesaurus*—a system for recognition and use of semantic information.

Living bodies are very dynamic internally. They exist on the basis of a complex net of controlled, interconnected transformations of matter and energy. The life of bodies is necessarily connected with regulated mechanisms of molecular interactions. First of all, it is the mechanisms of selective enzymatic catalysis of certain strictly defined chemical reactions under the control of genetic programs of NAs of genomes. This significantly increases the probability of highly improbable processes (including thermodynamically unfavorable processes), but ones that are absolutely necessary for the cell. Reactions are directed strictly to the required metabolic pathways, and billions of variants of unnecessary processes are automatically cut off.

Bodies, mechanisms, processes, and functions are the products of determined self-organization of matter into organic systems under the control of genetic information in the form of NAs and information from the environment. This results in the formation of ordered biological systems that exist on the basis of the principle of stable disequilibrium and support themselves by constantly working against increasing entropy. The creation and maintenance of order in living bodies, the implementation of all processes, and the use of information are necessarily accompanied by consumption, transformation, and targeted use of energy. Living bodies are capable of variation, adaptation, and survival, but exist only for a genetically limited period of time. The existence of any living body occupies just a moment in comparison with the billions of years during which the phenomenon of life has existed on Earth.

Organisms exist only as part of the ecosystem. Living bodies are autonomous, open, self-updating systems that interact with the environment through exchange of matter, energy, and information. They acquire the necessary substances and energy, renew the composition of the relevant molecules, and emit heat and chaos on the basis of genetic programs stored in DNA. This allows them to maintain their order and autonomy for a relatively long period of time.

Life originated, evolved, and exists today only on the basis of an aquatic environment, an internal "ether" that integrates everything in living bodies. The capacity of molecules for thermal motion constitutes the thermodynamic basis for the existence of living bodies. On the one hand, molecular motions condition all the different interactions, processes, and mechanisms that sustain life. On the other hand, they are ultimately the cause of destruction of the ordered state (entropy increase) and creation of structural instability. This internal conflict, or duality of synchronous destruction-recovery is one of the fundamental properties of living bodies. This non-equilibrium state is the condition and the driving force behind the continuous flow of controlled processes in living systems, directed against the increase of entropy, and toward the performance of useful work.

Any organism living today is derived from organisms that have lived before. Each is a link in an endless chain of evolution, which runs through billions of years to the present. And the chain continues to evolve rapidly towards an unpredictable future. Continuity in the existence of generations of living bodies is ensured by the presence of permanent genomes, which contain their genetic programs of reproduction, development, and survival. Genomes possess conservative variability, which on the one hand supports the continued existence of an unchanging form of phenotypes, and on the other hand, provides for the processes of their adaptation, development, and evolution. Cycles of copying (during the processes of reproduction and development) of NAs, proteins, and intracellular order, underlie the "intermittent continuity" of genomes and "continuous intermittence" of phenomes. Increasing entropy and resulting variability are spontaneous and inevitable. In order to remain stable and not disappear, living organisms must deal constantly with entropy, and this is only possible for a certain time. Therefore, for reliable and long-term existence, exact copies must be reproduced from time to time.

The life of organisms can be thought of as running on "in two dimensions". One is a hidden from our eyes: continuous genotypic life, a dynamic process of existence of a virtually unchanged genome. This is the essence of the life process. The other is the visible, intermittent phenotypic life: a periodic phenotypic expression of the genome in the form of living bodies. This is the concrete manifestation of life—the life of the genome in a phenotypic framework. From the perspective of a detached observer, life is an intermittent process of the existence of living bodies, whose task is survival and reproduction. In principle, that is the way it is. But organisms are just the carriers of a permanent genome; they ensure its survival and reproduction. And precisely for this purpose, living bodies carry a considerable baggage of resources for survival and reproduction, which we perceive as properties of the living. A carrier of life is recreated anew every time on the basis of NAs, which travel constantly from one body to another virtually without change. A body is one step in an infinitely repeating cycle of transformation and development of a discrete genome whose information it contains.

Thus, living bodies are phenotypic units of life, capable of self-organization and survival. The basis and reason for the existence of living bodies is the permanent genome. However, bodies are not the ultimate goal, but only a means for maintaining the phenomenon of life.

## 27.2 Primary Qualities of the Life Phenomenon

It is obvious that the molecules of DNA are the program and a means of structuring the surrounding material space into various forms of living organisms. A myriad of different DNA molecules contains a virtually infinite amount of genetic information. Functioning individually or in various combinations with other DNA in different genomes, these molecules specifically structure the surrounding material space, creating specific phenotypes in the form of individual representatives of millions of species of living organisms. Thus, we can say that the phenomenon of life is the continuous process of existence of an evolving system of genomes in various forms of physical (phenotypic) frameworks. We associate the phenomenon of life with the existence of the Integrated System of Life, which consists of the Global Genome and Global Phenome systems. These systems correspondingly consist of discrete genomes and phenomes. They function in an interrelated and inseparable manner, creating an integrated info-material field of life, which interconnects all living matter. This conception indicates the wholeness of life as a planetary phenomenon, despite its duality and discrete forms of manifestation.

Living bodies are the temporarily existing, dispensable products of the Global Genome at some stage of its development. This is the genome's habitat and medium for reproduction and evolution. Life is a process. This emphasizes the dynamism of the phenomenon and the meaning of various forms of its manifestation, indicating the successive changes in states and stages of development. It is a negentropy process, since it involves sequential accumulation of information and order in the integrated system of life. The ISL *evolves* continuously, remaining forever in a state of prolonged, progressive, and interconnected changes of genomes and living bodies, in accordance with certain laws of internal development and under the impact of external environmental factors. As a result, organisms become more complex and adaptable, which ensures homeostasis of genomes under virtually any conditions for millions of years. The evolution and eternity of life is based on the principle of consecutive mandatory replacement of its composing elements. That is, living bodies must cede their place in Nature to the following improved generations. Death and reproduction of organisms are necessary conditions for the life phenomenon. Cyclic recreation of genomes during reproduction and their continuation in offspring are absolutely necessary conditions for an infinite process. It becomes obvious that immortality of completed forms of individual organisms is impossible, since the changing generations of living bodies constitute the basis for evolution of the phenomenon of life, which ensures its permanence. Immortality of the life phenomenon also lies in its dual info-material nature. Despite the fact that the material components of living bodies wear out and die, the immaterial substratum of life in the form of information is not subject to ageing and death. Information is transferred from one body to another, and this conditions the recreation and continuity of life. Life is permanent and has been ever since the introduction of a highly organized system of interacting NAs and proteins in the colloidal medium in the form of networks of genomes and phenomes of cellular creatures. The ISL is steadily evolving and will inevitably come to an end when conditions on Earth are no longer compatible with the existence of aqueous solutions of NAs and proteins. The process of development of the Integrated System of Life on Earth is roughly halfway down this long road, and what we see today is only a tiny part of the great process of life.

## 27.3 Time and Life

The time during which particular living bodies exist is a mere moment compared with the infinite process of development of living matter. Time is one of the universal categories of the material world that determines the dynamics, orientation, and relationship of processes in space. It is a form of existence of matter which characterizes the duration of existence and the sequence and direction of changes of states of developing material systems. It is a measure of variability and is directed from the past into the future. Space is one of the universal characteristics of the material world that determines the length, structural properties, and coexistence and interaction of elements making up material systems.

Time and space are inseparable from matter, and hence from each other. In particular, in physics, the spatiotemporal characteristics of material bodies and systems possess four dimensions: three spatial and one temporal. Thus, any material body exists in a certain space and time for a specific period. This property is inherent to all living bodies.

The lifetime of an organism is determined genetically and varies greatly among different living beings (in our subjective understanding and sense of time): from a few minutes for single-celled organisms to thousands of years for some higher plants. Every organism is in a process of constant development. It appears at a certain time, develops for a certain time, reproduces for a certain period of time, gradually gets older, lives for a fixed time, and then dies.

In contrast to non-living bodies, living organisms can in a certain sense feel time and measure it. As "measuring units" they use natural cyclical phenomena or the duration of internal processes of their own bodies. Therefore, conventionally, we can talk about the standards of time which are sensed by living organisms. For example, day–night, spring–summer–autumn–winter, year, rainy season or drought, and others are exogenous standards, while the achievement of a certain size and shape, maturity, ability to move, generation of a sufficient amount of energy, duration of metabolic and physiological processes, etc., are endogenous standards of time. Sensing and measurement of endogenous and exogenous rhythms facilitates an adequate response and adaptation, orientating and synchronizing all the processes of life.

It is unlikely that even the higher animals, such as an ant, a crow, or a cat, are capable of a conscious sense of time. They are just genetically programmed to a specific development and behavior (connected with survival and reproduction) in certain circumstances of the outer and inner world. They are like robots, reacting only to their changes. They do not know how long they live or how long they will continue to live. They do not know about our hours and seconds, have no idea of the 4.5 billion years of Earth's existence, and do not feel either the vectors or the dynamics of the outer and inner processes. Thus, living organisms are just biological robots with a genetically determined "guarantee" concerning the duration of their existence. A conscious sense of time is perhaps inherent only to humans.

The duration of many metabolic and physiological processes in organisms (respiration, heart rate, reproduction, synthetic and energy cycles in cells, etc.) is conditioned by interactions between specific macromolecules, whose presence is determined by genetic programs. Thus, we can talk about an internal "genetic clock" which determines the duration of virtually all metabolic and physiological processes. So in fact the duration and direction of metabolic processes and functions, and hence the duration of life itself, is regulated by dosage and availability of information provided by a specific genome.

The different durations of endogenous processes in different organisms, and the various dependencies of living bodies on natural cyclic phenomena, leads to the same period of time being subjectively perceived as longer or shorter. Thus, despite the fact that astronomical time is the same for the whole Universe, it can have different quantitative measurements for different biological objects. In this case, the life cycle (LC) can be considered as a universal standard of measurement of biological time for different organisms. Despite the different durations of the LC in different organisms (from several hours to centuries), it is still a characteristic of the period of life, since it includes absolutely all the stages of individual development, from fertilization, birth, and development to reproduction, ageing, and death. From this perspective, all living organisms are considered under the same chronobiological conditions and live the same biological time—one life cycle. And the essence of this cycle is the development of a living body to sexual maturity, the production of gametes, which contain the immortal genome, and the transfer of this genome to future generations. All organisms (except humans) are fully satisfied with their lifetime. Each is allotted enough time for reproduction and development.

Another concept of biological time concerns the process of the life phenomenon (Fig. 27.1). This phenomenon appeared on Earth about 3.5 billion years ago and has existed continuously in time and space, despite the frequent change of generations of the carriers of life. It is obvious that the duration of the existence of any living body (the life cycle) is much shorter than the duration of the phenomenon of life. That is, the existence of an individual is only a flash (nothing) in comparison with the billions of years during which the phenomenon of life has existed on Earth. And the absolute duration of a life cycle has no fundamental significance for the existence of the life process as a phenomenon.

We have already noted that the global material system, which supports the phenomenon of life, is a set of evolving DNA molecules of all species of living organisms—the Global Genome. The time frames of existence of the evolving dynamic system of the GG are conditioned by cosmic processes. At the present time, the existence of this phenomenon is determined by the specific physical and chemical conditions on Earth, which are in turn conditioned by the Earth's special position in the Solar System. Given the temporal finiteness of the Solar System, we can conclude that the lifetime of the phenomenon of life is also finite.

But this period is defined by astronomical time frames on a cosmic scale, which is billions of years. Since the Sun is gradually "cooling", conditions similar to those on Earth could have existed several billion years ago on planets more distant

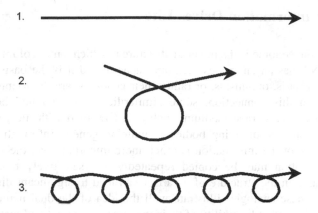

**Fig. 27.1** Integral vector of lifetime. 1—Time vector of the phenomenon of life: protracted, open, and endless under these conditions. It is not clear where it comes from and where it is heading. 2—Time vector of the lifetime of a single body: short, closed, and finite. It comes from the mother organism only for the purpose of reproducing genomes. 3—The integral vector of life on our planet is composed of an infinite number of cycles of living bodies which reproduce genomes

from the Sun. Thus, it is quite probable that the phenomenon of nucleic-protein life has already existed in the form of living bodies on other space carriers at a certain period of time. Changed physical and chemical conditions would have led to the disappearance of life's manifestations in such places, although the substances of life (NAs and proteins) could have survived in conditions of deep cold.

Thus, the Integrated System of Life from its very inception 3.5 billion years ago is in the process of irreversible development, in "the stream of evolution". This process continues today, every minute and every second, and will continue for as long as there is a Solar System, i.e., approximately another 5 billion years. We are roughly halfway down the road. We are a Global Phenome (the total system of all living creatures) which is characteristic only for this moment in the infinite development of the Earth. It is hard even to imagine what we will be in a few million years. Being within this process, being only an insignificant link in the chain of evolutionary transformations, it is very difficult to understand the essence, to determine the direction and goal of this movement.

Thus, time is a vector of movement of matter, the duration of movement of matter, and a measure of changes in the movement of matter. It is, probably, not even a physical, but, rather, a philosophical category, introduced by human thought for convenience of orientation in space. Space is also a philosophical category that was invented by man for comfort and directional orientation in the direction and range of motion of matter. For living systems, the time parameter is primarily an important evolutionary factor. Ever since its inception, the Integrated System of Life has been in a constant process of evolution which tends to infinity, while the sense of time only testifies to the finiteness of infinity.

## 27.4  How Information Drives Life

The DNA of the genome is the material structure in which a mass of information is recorded. DNA has an enormous memory capacity and it is the basis of genetic information. That is, organisms, or rather their genomes, are the living bearers of information. In this connection, some similarities may be noted between the properties of a data storage medium, such as a CD filled with files, and natural carriers, the genomes of living bodies, containing genetic information. In both cases, the value of the information is much more important than the value of the carrier. Information may be copied repeatedly and selectively read out. The slightest change in the structure of a carrier can lead to significant distortions of information. These analogies illustrate well the idea of the dual nature of living beings, the unity and indivisibility of their information content and matter, and the potential eternity of non-material information in comparison with any vulnerable material carrier.

Organisms are dissipating structures that are spontaneously destroyed under the impacts of thermal motion. In order to exist, living bodies must continually extract energy from the environment and use it in a targeted way to work against the forces of destruction. The direction of flows of energy and matter in certain ways and the regulation of the rates of these flows is carried out using the information of genomes. It is performed with special tools (proteins and enzymes) and mechanisms (means of interaction with other molecules). Initial, or initiating, information is contained in specific DNA of the genome. Realization of this information in the process of transcription provides the impetus for multiplication and cascaded dissemination of information. There appears a new type of information carrier— RNA. It represents a new potential and a new information content. In turn, this conditions a new round of multiplication, transformation, and distribution of information. New information carriers, called proteins, appear. They possess a colossal information potential and the physical and chemical basis for its application, leading to the emergence of the phenotypic manifestation of traits, or, we may say, the emergence of phenotypic information. That is, on the level of proteins, genetic information is transformed into phenotypic information. And the unit of genetic information, the gene, is transformed into the unit of phenotypic information, the protein. This is a new level of information in biosystems that conditions the appearance of myriad intracellular structures and an infinite variety of cells and multicellular organisms.

Matter exists in motion. Biological motion has a specific direction, rate, and extension. Objects can be made to move only by applying physical forces. Biological motion is controlled by the information of nucleic acids. In this case, biological information can be considered as a kind of natural force. We may assume that Nature is united by some informational force ether, where matter exists inseparably in interaction. This ether consists of many fields, systems, networks, and flows. One such field is the integrated space of circulation of biological information. This flow of information organizes a specific component of matter and controls it in a specific way. Thus, in Nature, there appears a biological form of matter.

In principle then, everything comes down to controlling flows of transformation of matter, energy, and information, and the main dualism of life is the inseparable coexistence of informational and material substances. Strategic importance is held by the genetic information that is implemented in living bodies through a multi-stage process of expression. Flows of matter and energy are directed strictly into defined paths, at a controlled rate, and into required directions, to organize matter into ordered structures.

The set of information of all discrete genomes forms the Integrated Informational Space of the ISL. Within it, there is a constant circulation of information maintained by special mechanisms. This circulation goes through particular organisms, which are the tools for implementing bio-information. Inside the GG, information circulates in the processes of replication, transcription, transgenesis, hybridization, and recombination. Inside the GP, informational flows are represented by the continuous intermittence of different living bodies in the processes of reproduction and alternation of generations. We comprehend the organized materialization of space and the manifestation of information in the form of specific living bodies as the phenomenon of life.

# Conclusion

Based on the foregoing, we can conclude that the phenomenon of life is manifested in the form of certain bodies. They are the bearers of life as a global phenomenon. The duality is clear, since on the one hand there are concrete physical bodies and on the other hand there is a property of Nature. Despite this, the phenomenon of life is unified, even though it is represented by numerous discrete units of living bodies—carriers of genomes. The genome is an integrating element between the phenomenon of life and living bodies. All genomes are united by the common nature of NAs into the integrated system of the Global Genome, and all organisms are united by the common nature of proteins into the integrated system of the Global Phenome. Being interdependent and interdetermining, the totality of genomes and phenomes forms the Integrated System of Life. The strategic aim of this organization of the ISL is the complementarity of the system of genomes with respect to diverse physical environments through qualitatively different phenotypes.

The phenomenon of life appeared on Earth as a system of genomes in discrete forms of phenotypic framework. Protobiont bodies could have emerged earlier, but it did not become a phenomenon until they began to interact, propagate, and disseminate. Then they formed a stable evolving system with fundamentally new properties. Evolution is a global property of the system of genomes, which is realized through the individual interactions of their material counterpart with the environment.

Living bodies are autonomous, but the Integrated System of Life is not autonomous. It is a network that covers the planet's surface. The bodies of living organisms consist of cells and the ISL consists of genomes in a phenotypic framework. The ability to reproduce, survive, age, and die is inherent in living bodies, but not in the phenomenon of life itself. Living bodies possess a finished form and composition, while the phenomenon of life is in the process of evolution. Organisms inevitably collapse and die, but reappear time and time again, while the Integrated System of Life arose many hundreds of millions of years ago, never gets old, does not die, and does not arise again, but is under constant development. Living bodies die, and even the species of genomes die, but the Global Genome

G. Zhegunov, *The Dual Nature of Life*, The Frontiers Collection,
DOI: 10.1007/978-3-642-30394-4, © Springer-Verlag Berlin Heidelberg 2012

and the phenomenon of life never perish. The death of individuals serves the purposes of the immortality of the life phenomenon by continuously updating the elements of the Integrated System of Life and ensuring the transition of genomes from one body to another. Time has a completely different meaning for living bodies and for the life phenomenon. A limit of existence is inherent in all organisms, but not the phenomenon of life. Science does not know of any example of an immortal body, while the Integrated System of Life has existed incessantly for billions of years. The existence of individuals is just a moment compared to the eternity of the life phenomenon. For permanent genomes, it does not matter how long their carriers exist.

Life as a phenomenon is linked to the planet Earth, just as the life of an organism is linked to a particular body. The phenomenon of life is like a perpetuum mobile of the second kind, a peculiar process which, once launched, continuously turns all the energy extracted from the environment into work. This machine has been working for several billion years, not only without losing, but even increasing its internal energy. It will continue to run for an infinite period, as long as the geophysical conditions for the existence of living bodies remain. Living bodies are the elements or, more accurately, the modules of this stable thermodynamic system of life. They are tools for direct extraction and transformation of energy, as well as binding mechanisms between the inanimate environment and the Integrated System of Life. It is living bodies that organize and direct the random thermal motion of the Universe into temporary maintenance of themselves and eternal maintenance of the system of life. They pay tribute to the laws of physics through their own deterioration and death, for the benefit of the eternity of the phenomenon of life.

Reproduction is the main property of living bodies, but not of the phenomenon of life, although eternity of the Integrated System of Life is conditioned by continuous procreation of its elements. Processes of adaptation, variation, heredity, and evolution of organisms all occur on the basis of reproduction and digenesis, which in turn is the foundation of the global evolution of the Integrated System of Life. In this view, living bodies are merely the means of survival and reproduction of genomes. The purpose of reproduction and evolution is not survival of organisms and species, but survival of genomes and the biological *information* they contain. The essence of life has not changed throughout billions of years. Only its phenotypic framework is changing, as in a kaleidoscope.

So in trying to answer the main question—what is life?—we come to certain conclusions that testify to the duality of this phenomenon. In essence, everything can be brought down to two points:

First of all, life can be represented as the obvious existence of a variety of *living bodies* and the objective existence of the *phenomenon of life.*

Secondly, the life of bodies and the phenomenon of life can be represented as the result of the interdependent co-existence and interaction of *informational and material substances of life.*

The duality of life is a part of the *paradigm of dualism* inherent in our world. For example, one can cite the following dualisms of Nature: substance and field of matter, kinetic and potential energy, dynamic and static forces, inertial and gravitational mass, and so on. In a deeper sense the duality implies indeterminacy.

Actually, trying to give a comprehensive definition of the phenomenon of life is a somewhat ungrateful task. Assuming that life is a qualitatively specific form of the existence of matter, we have to admit that it is just one of its many possible manifestations (e.g., a field, energy, and possibly other as yet unknown forms). But who can give an exact definition of matter? Who can answer the question of its origin? Are there any comprehensive definitions of energy, field, and the other basic categories of the material world? In each case, the answer is negative.

However, scientists have rather well studied and understood the essence of life and the diverse properties of living bodies. Numerous biotechnologies created by humans are evidence of our *precise knowledge* in this area. This is exemplified by the technology for cloning mammals, transgenic technology, gene engineering, methods of molecular diagnostics, methods of extracorporal fertilization, technologies of gene and cell therapy, as well as many other biotechnologies. They are very effective and are applied in hundreds of laboratories and clinics around the world, *demonstrating that we nevertheless have a sufficient understanding of life*.

# Recommended Literature

1. Mednikov, B.M.: The Axioms of Biology. Znaniye, Moscow (1982)
2. Trincher, K.S.: Biology and Information. Elements of Biological Thermo-dynamics. Nauka, Moscow (1965)
3. Bauer, E.S.: Theoretical Biology. VIEM, Leningrad (1935)
4. Alberts, B., Bray, D. et al.: Molecular Biology of the Cell: Garland Science, New York (1994)
5. Hopson J.L., Wessels N.K.: Essentials in Biology. McGraw-Hill Publishing Company, New York (1990)
6. Green, N., Stout, W., Taylor, D.: Biological Science. Cambridge University Press, Cambridge (1990)
7. Capra, F.: The Web of Life. Anchor Books, Doubleday, New York (1996)
8. Gilbert, S.: Developmental Biology. Sinauer Associates, Inc., Sunderland (1988)
9. Singer, M., Berg, P.: Genes and Genomes. University Science Books, California (1991)
10. Dawkins, R.: The Selfish Gene. Oxford University Press, Oxford (1976)
11. Dawkins, R.: The Extended Phenotype. Oxford University Press, Oxford (1982)
12. McClintock, B.: The significance of responses ofthegenome to challenge. Science **226**, 792–801 (1984)
13. Hadorn, E., Wehner, R.: Allgemeine Zoologie. Georg Thieme Verlag, New York (1977)
14. Joyce, G.F.: Directed molecular evolution. Sci. Am. **267**(6), 90−97 (1992)

# Index

**A**

Activity, 16, 68
Adaptation, 95, 96, 106
Aerobic respiration
Ageing, 123–125, 197, 266
Alleles, 83, 98, 229
Alternation, 85
Amphimixis, 87
Anabiosis, 4, 59, 61
Anabolism, 152, 167, 269
Animalia, 10
Anhydrobiosis, 58
Aqueous basis, 24
Aqueous medium, 24, 37, 123
Archaeobacteria, 243
ATP, 9, 13, 38, 48, 70, 76, 142, 143, 146–149,
    165–167, 169–171, 173, 175, 178, 181,
    182, 184, 202, 215, 269
Autonomy, 69

**B**

Bacteria, 8, 59
Biochemical reactions, 15, 18, 79
Biogenesis, 43
Bioinfogenesis, 251
Bioinformatics, 201
Biological creation, 148
Biological field, 221
Biological information, 201
Biological microprocessor
Biological oxidation, 147
Biosis, 57
Biosphere, 50
Biosystem, 53, 140
Blastomeres, 196

Bodily life, 261
Brownian motion, 24, 167

**C**

Catabolism, 167, 216
Catalysis, 141
Cell, 31, 139, 168, 177, 183
Cell cycle, 224
Cellular basis, 3, 25, 32
Chaos, 69, 119, 141
Chloroplast, 40, 79, 169
Chromatin, 87, 184, 186, 225
Chromosomal cycle, 226
Chromosomes, 225, 227, 84
Cloning, 151
Colloidal matrix, 27, 101, 270
Colonies, 11
Compartmentalization, 40, 79
Complementarity of life
Confined existence, 115
Copying, 82, 151, 153
Cryptobiosis, 58
Cybernetic system, 55, 202, 203
Cybernetics, 201, 202
Cytogenic processes, 66, 233, 251
Cytokenesis, 87
Cytoplasm, 8, 29, 154
Cytoplasmic matrix, 101
Cytosol, 76, 79, 143

**D**

Death, 23, 128, 129, 197, 273, 280
Development, 37, 195, 217, 273
Differential expression, 237, 272

G. Zhegunov, *The Dual Nature of Life*, The Frontiers Collection,
DOI: 10.1007/978-3-642-30394-4, © Springer-Verlag Berlin Heidelberg 2012

**D** (*cont.*)
Discreteness, 45, 48, 61, 183
Dissipative systems, 119, 153, 174
Diversity, 3, 11, 26, 214
DNA, 5–8, 11, 14, 16, 20, 22, 23, 28, 33, 35,
       38, 42, 43, 47, 48, 50, 57–59, 67, 75,
       77, 79, 80–85, 87, 93, 97, 100, 101,
       103, 105, 106, 108–110, 115, 117, 120,
       124, 125, 128, 130, 146–148, 151, 152,
       174, 184–186, 203, 205, 207, 208,
       211–215, 217–225, 227, 229–233,
       235–240, 242, 244, 245, 247–250, 252,
       254, 255, 261, 263–265, 268–271, 273,
       274, 277–279, 282, 284
Dualism, 265, 266, 274
Duality of life, 61, 131, 261


**E**
Ecosystems, 50
Energetic basis, 21
Energy, 20, 48, 51, 165, 168, 169, 175, 188,
       208
Enthalpy, 175
Entropy, 115, 140
Enzyme, 23, 143
Episome, 31, 272
Eukaryotes, 7, 10, 30, 40
Evolution, 42, 61, 85, 107, 111, 267
Exon, 235
Expediency, 12
Expression, 110, 220, 233


**F**
Fertilization, 196
Functional systems, 188, 241
Fungi, 9, 166


**G**
Gametes, 85, 86, 266
Gametogenesis, 195
Gene, 208, 233, 234, 244
Gene expression, 208, 233, 234
Generation, 5, 71, 83, 111, 155
Generative cells, 133
Gene transfer, 244
Genetic apparatus, 6, 76, 100, 107
Genetic code, 223
Genetic continuity, 251
Genetic homeostasis, 232
Genetic individuality, 230, 231, 232
Genetic information, 22, 27, 33, 216, 219, 221

Genetic material, 224, 225
Genetic network, 235
Genetic program, 57, 58, 97, 211, 214
Genetic systems, 235, 243
Genome, 12, 32, 60, 71, 84, 96, 108, 110, 119,
       131, 132, 208, 218, 222, 224, 237, 244,
       246, 247, 252, 264, 267, 268, 280, 282
Genotype, 153
Genotypic evolution
Genotypic life, 265, 266, 267
Global Genome (GG), 6
Global Phenome (GP), 6, 264


**H**
Hereditary information, 48
Heredity, 85, 96, 100, 107, 232
Hierarchy, 46
Homeostasis, 113, 115
Homogenesis, 43
Hypobiosis, 58
Human genome, 238


**I**
Immortality, 273, 280
Individual, 89, 107, 131, 132
Individual development, 89
Infobiogenesis, 252
Info-genetic continuity, 253
Info-genetic mechanisms, 217, 218
Info-genetic process, 217
Information, 33–35, 201, 202, 204, 207–210,
       214, 219, 230, 252, 255, 280, 284
Integrated information space, 30, 247
Integrated Life System
       (ILS), 131
Integrity, 45, 222
Intelligent life, 261
Interactions, 176, 187, 220
Intron, 235
Intragenomic information, 35
Intergenomic information, 35
IS-elements, 105, 245


**K**
Karyoplasm, 248, 76, 270
Karyotype, 23, 231, 269


**L**
Latent life, 5, 57, 269
Levels of life, 48, 107

Life, 3, 5, 14, 15, 17, 19, 21, 23, 25, 28, 37, 42, 43, 46, 48, 57, 58, 60, 65, 108, 131, 132, 175, 176, 210, 217, 218, 252, 255, 264, 266, 274, 277–280, 281, 283, 284
Life carries, 115
Life cycle (LC), 128
Life development, 37, 39
Life manifestation, 18, 59
Life origin, 278
Life states, 57
Life stopping
Living bodies, 60, 67
Living computers, 212, 216
Living system, 55, 69, 172

**M**
Macromolecules, 37
Macroprocesses, 183, 186, 187
Matrix processes, 153
Matter and energy exchange, 66
Mechanisms, 43, 47, 248
Membrane, 7, 20, 29
Memory, 204, 213
Metabolism, 49, 166, 191
Metabolic chains, 143, 211
Mitochondria, 76, 183, 171
Meiosis, 17, 35, 82, 251
Mitosis, 184, 266
Monera, 75
Movement, 68
Multicellular organism, 50, 68, 70, 74, 75, 85, 129, 133, 195, 209
Mutation, 105

**N**
Natural selection, 103, 106, 118, 153, 154, 245, 252, 275
Negentropy, 76, 93, 210, 280
Non-cellular form of life, 28
Nonequilibrium system, 172
Nucleic acid (NA), 21
Nucleotide, 232
Nucleotide sequences, 38, 152, 229
Nucleus, 53, 77

**O**
Offspring, 83, 102, 153
Open system, 203
Order, 45, 47
Organelles, 48, 143, 208

Organism, 46, 49
Organization, 45

**P**
Parthenogenesis, 86, 90
Phage, 30, 35
Phenome, 13, 60, 110, 131, 218, 224, 237, 244, 247, 264, 267, 280, 283
Phenomenon of life, 6, 18, 41, 280
Phenotype, 6, 29, 90
Phenotypic diversity, 6
Phenotypic evolution, 267
Phenotypic framework, 23, 28, 60
Phenotypic life, 6, 267, 274
Photosynthesis, 181
Phylogenesis, 37, 90, 220, 237
Physiological processes, 191
Plantae, 9
Plasmid, 29, 35, 105, 240
Principels of organization, 45, 240
Probability of life, 17
Process, 5, 6, 17
Prokaryotes, 8, 76, 234
Pronucleus, 87
Proteins, 23, 50, 78, 116, 150, 152, 165, 189, 193, 208, 266, 269, 277
Protista, 8, 9, 75
Proteome, 23
Protobionts, 38, 42, 61, 152

**Q**
Quantum mechanisms, 179

**R**
Reparation, 152
Replication, 84, 152, 184
Reproduction, 11, 81, 85, 131, 264
Respiration, 21, 57, 60, 74
Ribozymes, 38, 144
RNA, 6, 7, 11, 22, 28, 29, 33, 38, 42, 47, 48, 50, 75, 79, 87, 92, 108, 115, 124, 144, 151–153, 184, 205, 207, 213–215, 219, 220, 221, 222, 229, 230, 233, 235, 239, 249, 252, 268, 277, 284

**S**
Selection, 154
Self-maintenance, 113
Self-organization, 212, 217

**S** (*cont.*)
Self-preservation, 45, 46, 77, 135
Self-reproduction, 38, 67, 176
Signals, 204
Similarity, 12
Space, 281, 283, 285
Somatic cells, 83, 85, 86, 99, 122
Species, 34, 35, 49, 50, 81, 122, 230, 231, 232, 239, 240, 268
Stem cells, 121
Stem molecules, 22
Substrates of life, 18, 273
Survival, 14, 33, 65, 209
Symplast
Syncytia
Syngenesis, 67

**T**
Temperature of life, 207
Thermodynamic basis, 278
Thesaurus, 201
Time and life, 281
Traits, 229, 230, 231–233, 237, 240
Transcription, 152, 220, 229, 233, 234, 235, 236, 239, 240

Transgenesis, 230, 232, 233, 244, 247, 252
Translation, 153, 220, 234, 236
Transmission of information, 33
Transposon, 110

**U**
Unicellular organism, 4, 75, 254
Universality, 12, 222, 223
Unlikely events, 17

**V**
Vertebrates, 41, 49
Viroid, 29, 30, 105, 245
Virus, 7, 29, 65, 272

**W**
Water, 4, 24, 25, 240

**Z**
Zygote, 233, 237

# Titles in this Series

**Quantum Mechanics and Gravity**
By Mendel Sachs

**Quantum-Classical Correspondence**
Dynamical Quantization and the Classical Limit
By Josef Bolitschek

**Knowledge and the World: Challenges Beyond the Science Wars**
Ed. by M. Carrier, J. Roggenhofer, G. Kuüppers and P. Blanchard

**Quantum-Classical Analogies**
By Daniela Dragoman and Mircea Dragoman

**Life - As a Matter of Fat**
The Emerging Science of Lipidomics
By Ole G. Mouritsen

**Quo Vadis Quantum Mechanics?**
Ed. by Avshalom C. Elitzur, Shahar Dolev and Nancy Kolenda

**Information and Its Role in Nature**
By Juan G. Roederer

**Extreme Events in Nature and Society**
Ed. by Sergio Albeverio, Volker Jentsch and Holger Kantz

**The Thermodynamic Machinery of Life**
By Michal Kurzynski

**Weak Links**
The Universal Key to the Stability of Networks and Complex Systems
By Csermely Peter

G. Zhegunov, *The Dual Nature of Life*, The Frontiers Collection,
DOI: 10.1007/978-3-642-30394-4, © Springer-Verlag Berlin Heidelberg 2012

296                Titles in this Series

**The Emerging Physics of Consciousness**
Ed. by Jack A. Tuszynski

**Quantum Mechanics at the Crossroads**
New Perspectives from History, Philosophy and Physics
Ed. by James Evans and Alan S. Thorndike

**Mind, Matter and the Implicate Order**
By Paavo T. I. Pylkkänen

**Particle Metaphysics**
A Critical Account of Subatomic Reality
By Brigitte Falkenburg

**The Physical Basis of The Direction of Time**
By H. Dieter Zeh

**Asymmetry: The Foundation of Information**
By Scott J. Muller

**Decoherence and the Quantum-To-Classical Transition**
By Maximilian A. Schlosshauer

**The Nonlinear Universe**
Chaos, Emergence, Life
By Alwyn C. Scott

**Quantum Superposition**
Counterintuitive Consequences of Coherence, Entanglement, and Interference
By Mark P. Silverman

**Symmetry Rules**
How Science and Nature Are Founded on Symmetry
By Joseph Rosen

**Mind, Matter and Quantum Mechanics**
By Henry P. Stapp

**Entanglement, Information, and the Interpretation of Quantum Mechanics**
By Gregg Jaeger

**Relativity and the Nature of Spacetime**
By Vesselin Petkov

**The Biological Evolution of Religious Mind and Behavior**
Ed. by Eckart Voland and Wulf Schiefenhövel

**Homo Novus - A Human Without Illusions**
Ed. by Ulrich J. Frey, Charlotte Störmer and Kai P. Willführ

## Brain-Computer Interfaces
Revolutionizing Human-Computer Interaction
Ed. by Bernhard Graimann, Brendan Allison and Gert Pfurtscheller

## Extreme States of Matter
on Earth and in the Cosmos
By Vladimir E. Fortov

## Searching for Extraterrestrial Intelligence
SETI Past, Present, and Future
Ed. by H. Paul Shuch

## Essential Building Blocks of Human Nature
Ed. by Ulrich J. Frey, Charlotte Störmer and Kai P. Willführ

## Mindful Universe
Quantum Mechanics and the Participating Observer
By Henry P. Stapp

## Principles of Evolution
From the Planck Epoch to Complex Multicellular Life
Ed. by Hildegard Meyer-Ortmanns and Stefan Thurner

## The Second Law of Economics
Energy, Entropy, and the Origins of Wealth
By Reiner Kümmel

## States of Consciousness
Experimental Insights into Meditation, Waking, Sleep and Dreams
Ed. by Dean Cvetkovic and Irena Cosic

## Elegance and Enigma
The Quantum Interviews
Ed. by Maximilian Schlosshauer

## Humans on Earth
From Origins to Possible Futures
By Filipe Duarte Santos

## Evolution 2.0
Implications of Darwinism in Philosophy and the Social and Natural Sciences
Éd. by Martin Brinkworth and Friedel Weinert

## Probability in Physics
Ed. by Yemima Ben-Menahem and Meir Hemmo

**Chips 2020**
A Guide to the Future of Nanoelectronics
Ed. by Bernd Hoefflinger

**From the Web to the Grid and Beyond**
Computing Paradigms Driven by High-Energy Physics
by René Brun, Federico Carminati and Giuliana Galli Carminati

**Natural Fabrications**
Science, Emergence and Consciousness
By William Seager